高等职业教育农业农村部“十三五”规划教材

畜牧兽医专业英语

卫金珍◎主编

Specialized English in Animal Husbandry and Veterinary Medicine

中国农业出版社
北　　京

内容简介

《畜牧兽医专业英语》是专门针对畜牧兽医专业的高职学生和具有一定英语基础的中职学生编写的。

全书共有 10 个单元，分别围绕畜牧兽医的 10 个重要主题：畜牧业经济、畜禽品种、解剖基础、遗传育种、牧场建设、科学饲养、疾病防治、畜牧兽医公共卫生、宠物和畜产品。每个单元都包括 Warming up，Listening and Speaking，Reading，Grammar，Extra Reading，Writing Practice 和 Vocabulary 部分。Warming up 部分主要是用与主题相关的图片导入学习；Listening and Speaking 部分主要是围绕相关主题，结合实际，以对话的形式学习；Reading 部分是最能体现主题的畜牧兽医专业知识的短文；本教材的 Grammar 部分重点讲解英语的句子结构，句子成分和科技英语的写作要点；Extra Reading 部分有两篇与主题有关的短文，供读者拓展学习；Writing Practice 部分为读者设计了与本单元有关的写作练习。

本教材配有包括听说、阅读及词汇部分的听力材料(扫描文中二维码即可免费收听)，也为老师们上课准备了 PPT(可登录中国农业教育在线 www.ccapedu.com 查询使用)。

编审人员

主　编　卫金珍

副主编　李小宁　陈士华

编　者　（以姓名笔画为序）

卫金珍　李小宁　李朝波

张　蕾　陈士华　郭严如

语言主审　Steven Pope

专业主审　杨再昌

序

我很少为教材写序，因为懒惰和本人的水平不太理想，偶尔给几位作者的大作写过序，皆因其水平高上。贵州农业职业学院卫金珍老师托我为其主编的、由中国农业出版社出版的高等职业教育农业农村部“十三五”规划教材《畜牧兽医专业英语》写个序，我就照办了，因为盛情难却，也因为我愿意写这个序，以此表达我由衷的钦佩之情。

我初识卫老师是在贵州省高等职业技能大赛的比赛现场，当时她带领贵州农业职业学院比赛队参赛，之后的比赛，她一直坚持带队参加。我作为大赛的裁判长，见证了贵州省各职业院校参赛队的高超的职业技能水平，这与老师们的精心教学和悉心指导是分不开的。卫老师在贵州农业职业学院教授英语课程，她对职业英语教学和研究的投入与执着给我留下了深刻的印象。当她把多年职业英语教学和实践的结晶——《畜牧兽医专业英语》的书稿呈现在我面前时，我十分高兴。一本书几易其稿，反复斟酌，写了数年，如今终得出版，多年的辛勤耕耘有了结果，值得庆贺和宣扬。

这本教材一共10个单元，每个单元均由准备活动（Warming up）、听说技巧（Listening and Speaking）、阅读理解（Reading）、语法知识（Grammar）、补充阅读（Extra Reading）和写作实践（Writing Practice）等几部分组成，内容涉及畜牧业经济、畜禽品种、解剖基础等10个畜牧兽医主题。其使用对象主要为职业院校畜牧兽医专业的学生，编写目标明确，注重实用性和时代性，内容紧紧与畜牧兽医专业知识结合，选材贴近职业，贴近学生，贴近生活。教材的内容设计版式新颖，图文并茂，直观性强。

改革开放40年来，职业教育历经从计划经济到市场经济、从行业办到社会办、从理论学科型教学到真正意义上的实训型职业教育等数次转型。与此同时，我国中等职业教育累计培养了8 000多万名毕业生，高等职业教育培养了2 000多万名毕业生，职业教育共为国家输送了1亿多名高素质劳动者和技能型人才。近年来，我国职业教育又有了前所未有的发展，对外交流越来越广泛，各行各业急需大量具有国际视野以及具有较高专门技能的科技人才。

职业教育专业类别繁多，每个专业类别都需要有相应的教材，以助培养有较高专门技能的对外交流科技人才。由卫老师主编的《畜牧兽医专业英语》正是顺应了这个发展趋势，值得祝贺。该教材针对畜牧兽医专业的高职学生和具有一定英语基础的中职学生编写，紧密结合教学实践，融思想性、趣味性、可读性和实用性于一体，其英语语言规范、篇幅适中、由浅入深、循序渐进，既利于教学，也便于学生自学，实属难得的教学与自学的好教材。

是为序。

李炳林
贵州大学外国语学院教授
中国英语写作教学研究会副会长
中国认知神经语言学研究会常务理事
2018年3月于贵州大学外语楼117研究室

前言

近年来，我国职业教育有了前所未有的发展，对外交流越来越广泛，各行各业急需大量具有国际视野以及具有较高专门技能的科技人才，能够熟练运用畜牧兽医专业英语的人才的需求量也不断增加。根据目前职业教育和畜牧业快速发展的需要，《畜牧兽医专业英语》教材应运而生，本教材由贵州农业职业学院具有畜牧兽医专业知识并多年从事职业教育的英语高级讲师卫金珍担任主编。广西农业职业技术学院的李小宁担任副主编，参与编写的人员还有辽宁职业学院的陈士华，江苏农牧科技职业学院的张蕾和贵州农业职业学院的郭严如、李朝波。所有编写人员都是来自我国职业院校的英语教师或者从事畜牧兽医的专业教师。本教材特聘贵州大学杨再昌教授作为专业主审；美国专家Steven Pope博士作为语言主审。听力词汇、对话、阅读材料和课文生词由美国志愿者Hugo、贵州农业职业学院英语教师陆晔和易如朗读。录音之前，志愿者Hugo、陆晔老师和易如老师对稿件作了认真细致的校对工作，并提出了宝贵的修改建议。贵州农业职业学院翁玲老师提供了许多与专业相关的精美图片，在此一并表示感谢。

本教材的主要适用对象是我国高等职业院校畜牧兽医专业的学生、畜牧兽医从业人员以及一部分英语基础较好的中等职业学校的学生，是一本普通英语和大学畜牧兽医专业英语的衔接教程。教材编写注重实用性和时代性。编写内容紧紧与畜牧兽医专业知识结合，选材贴近职业，贴近学生，贴近生活。为了便于教学，本教材专门配有同步的PPT课件。

本教材的具体分工为：Unit 1（郭严如），Unit 2、Unit 3（卫金珍），Unit 4（陈士华、张蕾），Unit 5（张蕾），Unit 6（陈士华），Unit 7、Unit 8（李小宁），Unit 9（李朝波），Unit 10（郭严如、李朝波）。其中李小宁、陈士华和郭严如三位老师承担了部分统稿任务。

本教材的顺利完成，得益于有关专家的指点和帮助，在此表示衷心的感谢！

由于编者水平有限，教材中难免有疏漏和不足之处，恳请各位读者多提宝贵意见。

编　者

2018年7月

Contents

序
前言

Unit 1 Livestock Husbandary Economy ········· 1
Warming up ········· 1
Listening and Speaking ········· 2
Reading ········· 4
Grammar ········· 6
Extra Reading ········· 8
Writing Practice ········· 11
Vocabulary ········· 11

Unit 2 Breeds of Livestock and Poultry ········· 13
Warming up ········· 13
Listening and Speaking ········· 14
Reading ········· 15
Grammar ········· 17
Extra Reading ········· 19
Writing Practice ········· 23
Vocabulary ········· 23

Unit 3 Basic Knowledge of Anatomy ········· 25
Warming up ········· 25
Listening and Speaking ········· 26
Reading ········· 28
Grammar ········· 30
Extra Reading ········· 32
Writing Practice ········· 35
Vocabulary ········· 36

Unit 4 Genetics and Breeding …… 38

Warming up …… 39
Listening and Speaking …… 39
Reading …… 41
Grammar …… 42
Extra Reading …… 44
Writing Practice …… 47
Vocabulary …… 48

Unit 5 Pasture Construction …… 49

Warming up …… 49
Listening and Speaking …… 50
Reading …… 52
Grammar …… 53
Extra Reading …… 55
Writing Practice …… 58
Vocabulary …… 59

Unit 6 Scientific Feeding …… 60

Warming up …… 60
Listening and Speaking …… 61
Reading …… 62
Grammar …… 65
Extra Reading …… 67
Writing Practice …… 70
Vocabulary …… 70

Unit 7 Disease Prevention and Control …… 72

Warming up …… 73
Listening and Speaking …… 74
Reading …… 75
Grammar …… 76
Extra Reading …… 79
Writing Practice …… 82

Vocabulary ········· 82

Unit 8 Veterinary Public Health ········· 84

Warming up ········· 84
Listening and Speaking ········· 85
Reading ········· 87
Grammar ········· 88
Extra Reading ········· 90
Writing Practice ········· 93
Vocabulary ········· 94

Unit 9 Pets ········· 95

Warming up ········· 95
Listening and Speaking ········· 96
Reading ········· 98
Grammar ········· 100
Extra Reading ········· 102
Writing Practice ········· 105
Vocabulary ········· 106

Unit 10 Animal products ········· 108

Warming up ········· 108
Listening and Speaking ········· 110
Reading ········· 111
Grammar ········· 113
Extra Reading ········· 115
Writing Practice ········· 118
Vocabulary ········· 118

参考答案 ········· 120
参考文献 ········· 161

Unit 1

Livestock Husbandary Economy

Upon completing this unit, you will be able to

◆ *identify and understand the key English words or terms related to the animal husbandry economy;*

◆ *master the basic elements of English sentences;*

◆ *communicate in relevant activities.*

Warming up

1. Which animals appear to feel the most comfortable in the below pictures? Why?

____

____

2. Where would you raise animals to best maximize profitability? Please provide your reasoning.

____

____

Words and Expressions

husbandry [ˈhʌzbəndrɪ]
n. 饲养；耕种

veterinary [ˈvetrənərɪ]
adj. 兽医的　n. 兽医（等于 veterinarian）

ecological agriculture　生态农业

Holstein [ˈhɒlstaɪn]
n. 荷兰的一种乳牛；黑白花牛

dairy cattle　奶牛

Part A　Complete the following tasks according to the instructions.

1. Listen to the recording and then judge whether the following statements are true (T) or false (F).

(1) She is employed at a university. ( )

(2) He is her old friend. ( )

(3) Emi lives in Apartment # 23. ( )

(4) Guo is majoring in animal husbandry and veterinary. ( )

(5) Both of them come from the same university. ( )

2. Listen to the dialogue again until you can write down what the speakers say.

Emi: ______________________________

Guo: ______________________________

Emi: ______________________________

Guo: ______________________________

Emi: ______________________________

Guo: ______________________________

Emi: ______________________________

Guo: ______________________________

Emi: ______________________________

3. Work in pairs. Find a partner, then tell her or him about the schoolmate you met on the first day that you arrived at your college.

Part B Finish the following tasks.

1. Guo is in the classroom. Listen and check the right one.

Guo is studying ________.

A. Cattle B. Sheep C. Person

2. Listen again. Guo is looking for a picture. What does the picture look like? Find the right answer.

A. It's tall / short.

B. It's yellow / black and white.

C. It has long / short hair.

D. Milk / Wool comes from these cattle.

3. Listen again and choose the right picture. ________

A

B

C

Animal Husbandry Economy

As one major part of agriculture and one of the important material-producing departments, animal husbandry, based on the production capacity of flora and fauna, has been creating economic value by breeding livestock in captivity to obtain animal product or working animals. In husbandry, livestock comes in two types: the farming animals for meat like chicken, horses, cows and so on, and commercial livestock like deer, bear, marten etc. Modern husbandry not only provides meat, eggs, milk and blood for human beings but also supplies raw material like fur, leather and bones to light industry. Meanwhile, animal husbandry offers invaluable assistance to farming through the aspects of manure and working animals.

Throughout the world, different countries adopt various economic modes to develop animal husbandry on basis of different features of land, capital and labor. For example, the Australia-New Zealand Mode refers to grassland input being the main speciality; the Europe Intensive Mode relies mainly on the capital input; while the Traditional Mode's main characteristic is labor input.

China used to engage in the Traditional Mode to develop animal husbandry. However, the impact of shrinking areas of grassland, less farming land for a large population, and fewer farms for huge numbers of farmers has meant that during the last few decades Chinese animal husbandry has been experiencing the transition from traditional husbandry to modernized husbandry.

Owing to poor quality of grassland resources, great areal variation and unbalanced economic development, Chinese husbandry has gradually accommodated a diversity of forms. These forms include: individual farmers, professional farmers, big farm and large-scale feeding areas. On the whole, farmer's family management is still the main form of the husbandry production. Through the continued process of industrialization and urbanization in China, increasingly the form of the individual farmer develops into the big farm. On basis of the characteristics of Chinese animal husbandry and its steady development into a more European mode, Chinese animal husbandry is developing into a scale operation and will gradually realize its modernization in the near future.

Your tasks

1. Write down all the words you do not know in the passage. Then check out the pronunciations and read them aloud.

____

____

2. Understanding the text: answer the following questions with the information contained in the reading.

(1) According to the text, what is animal husbandry? Briefly define the economy of animal husbandry.

____

(2) What are the three modes of animal husbandry?

____

(3) What are the differences between the Europe Intensive Mode and Traditional Mode of agriculture?

____

(4) Which economic mode did China use and which will be used in the future? Why?

____

(5) In your opinion, will Chinese animal husbandry economy become more successful in the future?

____

3. Fill in the blanks with the words and expressions given below in correct forms. Each one can be used only once.

by-product refer to adopt variation industrialization

(1) Miss Wang's life is full of change and ________ .

(2) Meat, wool and leather are all ________ obtained in the process of animal husbandry economy.

(3) ________ is the core of Chinese economy.

(4) The foreign couple ________ a disabled baby in orphanage.

(5) Mary never ________ her parents in her self-introduction.

Grammar

句 子 成 分

句子成分是构成句子的各个部分。在英语中，它主要包括：主语、谓语、宾语、表语、定语、状语、主语补足语和宾语补足语等，主要由实词（名词、动词、代词、形容词、副词和数词）充当。虚词（介词、连词、冠词和感叹词）不能单独充当句子成分，它在句子中仅起连接、限定、表达内心情感等作用。

句子成分	功　能	相应的词或短语	例　句
主语	句子的中心和动作的出发者或承受者	能充当主语的有：名词、代词、数词、动名词或动名词短语、不定式等	Everything is impossible. To be a brave man requires great willpower.
谓语	用来说明主语的动作或状态	谓语只能用动词担当。行为动词和be动词可以单独作为谓语；情态动词、助动词不能单独作为谓语	He wants to look for a beautiful and kind girlfriend. We will raise 500 pigs this year.
表语	用来说明主语的身份、性质、特征、内容、数量等	能充当表语的有：名词、代词、形容词、副词、动词不定式、动名词、数词	He is 18 years old. Your English teacher is handsome. My dream is to be a doctor.
宾语	谓语动词动作的承受者。有直接、间接和复合宾语之分	能充当宾语的有：名词、代词、数词、动词不定式、动名词	That girl likes dogs. You give me a present. A bad mood makes things confused.
定语	有前置和后置定语两种，它是用来修饰名词或代词的	能充当定语的有：形容词、代词、数词、名词、名词所有格、分词、动名词、动词不定式、介词短语	Curled hair looks more beautiful. The student in the first line is our monitor. Do you want anything else?
状语	用来修饰动词、形容词、副词或整个句子	由副词、分词、介词短语和不定式充当	After 20 years Lucy will be a CEO. I love music very much.

Exercises

1. Tell the sentence elements in the following underlined parts.

(1) To be a farmer was the happiest thing in her mind.

(2) Those students should be praised.

(3) Although Lucy is angry, she is still talking with her friend in a soft voice.

(4) Anyone who wants to be first can tell you how hard it is.

(5) I gave her my gift.

2. Analyze the underlined sentences in the following passage and then translate the paragraph into Chinese.

With the development of the Chinese economy, every industry has been fumbling for a right way to confront the serious challenges facing China. As for the development of Chinese animal husbandry, two obvious developing models have existed: one is grassland animal husbandry and animal husbandry in mountainous regions on natural grassland, and the other is shed animal husbandry. According to Chinese natural conditions, grassland animal husbandry and animal husbandry in mountainous region on the natural grassland have been the main parts of Chinese animal husbandry. With the development of science and technology in recent years, shed animal husbandry is thriving.

3. Translate the following sentences into English.

(1) 薯条是土豆的副产品之一。(by-product)

(2) 这篇报道中从未提到过中国畜牧业经济的发展状况。(refer to)

(3) 我们农场采用了新的饲养技术以适应社会的发展。(adopt to)

(4) 我国畜牧业的特色是什么?(speciality)

(5) 畜牧业发展日益繁荣昌盛。(thriving)

There are two passages followed by five questions respectively. Read them and then circle the correct answer A, B, C or D for questions 1-10.

Grassland animal husbandry, one of the circular economic modes, obtains animal by-products by breeding horses, sheep, cows and other animals with the natural grassland being the main resource. According to historical research, grassland animal husbandry can be traced back to mid-to-late primitive society. During this period of time, human beings came to understand the behaviour and the possibilities of animals being domesticated while hunting animals and coexisting with them in a hostile state of nature. The domestication of animals evolved from breeding smaller animals, such as dogs, pigs and chickens, to breeding larger animals, such as goats, antelopes and horses. Animal husbandry eventually thrived through years of research and practice. Because of its reliance on natural grasslands, grassland animal husbandry is forced to be constantly moving with the change of seasons, and has the obvious characteristics of being at the mercy of the nature in regards to feeding livestock and living where there are water and grass.

Compared with the large-scale farming, animal husbandry in mountainous regions is generally on a smaller-scale and is relatively more stable. This is because the production activities of animal husbandry in mountainous regions are engaged in the uncultivable slopes area. So, its grazing areas are stable and the types of bred animals are limited (e. g. cattle and sheep). Animal husbandry in mountainous regions always exists in the farming or semi-farming areas, where human beings engage in husbandry production activities on the uncultivable slopes to increase income. According to historic data, animal husbandry in the mountainous regions is much later than grassland animal husbandry.

1. About grassland animal husbandry, all of the following are correct except that ________ .

A. it can be used to breed sheep, horses, cows and other domesticated animals

B. natural grasslands are a good resource for animals

C. it is unstable and changing

D. water and grass are basic elements for grassland animal husbandry

2. According to research, grassland animal husbandry can be traced back to the mid-to-late primitive society. "traced back to" means ________ .

A. follow or discover

B. find the origin of something

C. copy

D. sketch the outline of something

3. Animal husbandry in mountainous region is all but ________ .

A. small-scaled

B. relatively stable

C. in farming or semi-farming areas

D. in the cultivable land with grass

4. According to the passage, grassland animal husbandry tends to ________ than animal husbandry in mountainous region.

A. be more stable

B. be large-scaled

C. emphasize more on historic data

D. emphasize less on animal by-product

5. The best title for the text is ________ .

A. a brief introduction to animal husbandry in mountainous regions in China

B. a brief introduction to grassland animal husbandry in China

C. the contrary of grassland animal husbandry and animal husbandry in mountainous regions in China

D. the differences between grassland animal husbandry and animal husbandry in mountainous regions in China

Shed animal husbandry, also known as livestock breeding, is a human activity of intensively breeding livestock by building sheds and enclosing lands to build farms. According to the species of livestock, this can include livestock fenced into horse farms, cow farms, chicken farms etc.

Of course, there are some comprehensive farms, in which more than one species of livestock are bred. Shed animal husbandry is a much later development than the above two modes. The reason for this is that livestock breeding requires a higher level of science-and-technology development and management because livestock breeding is relatively stable and intensive. It also needs the support of the specialized veterinary science, especially in breeding animals like foxes, bears, deer and so on. In addition, shed animal husbandry relies greatly on large capital investment. In recent years, with the accelerated development of science and technology in China, the improvement of economic power and support from government, livestock breeding is thriving and fast developing.

6. The best title for the passage is ________.

A. the advantages of shed animal husbandry in China

B. the current situation of shed animal husbandry

C. a brief introduction to shed animal husbandry

D. the growing trend of shed animal husbandry in China

7. The word "enclosing" in the first sentence of the passage most likely means ________.

A. putting in

B. surrounding completely

C. closing in or confining

D. enveloping

8. The development of shed animal husbandry occurred much later than other forms of husbandry, which one is the incorrect reason in the following answers? ________

A. Higher level for science and technology is needed

B. Centralized management is not needed

C. It needs financial support

D. It needs the support from the veterinary science

9. According to the passage, we can see that the attitude of the author is ________.

A. objective

B. subjective

C. neutral

D. worried

10. The passage shows that ________ .

A. it is important for people to use shed animal husbandry as an only way to breed animals

B. shed animal husbandry has been popular since ancient times

C. shed animal husbandry can be easily and naturally developed

D. shed animal husbandry will enter a new rapid development stage in the future

Writing Practice

Imagine that you are a freshman majoring in animal husbandry and veterinary science. For your safety, the college student assistant needs you to fill in a personal information form as soon as you come to the college. Please finish the form based on your own information.

Personal Information Form

<table>
<tr><td>Name</td><td></td><td>Gender</td><td></td><td>Date of birth</td><td colspan="2"></td><td rowspan="4">Photo</td></tr>
<tr><td>Student ID</td><td></td><td>Nationality</td><td></td><td>Political status</td><td colspan="2"></td></tr>
<tr><td>Department</td><td></td><td>State of health</td><td></td><td>Major</td><td colspan="2"></td></tr>
<tr><td>Address</td><td colspan="6"></td></tr>
<tr><td>Graduate school</td><td colspan="2"></td><td>Hobby</td><td colspan="4"></td></tr>
<tr><td colspan="2">Phone NO.</td><td colspan="2"></td><td colspan="2">E-mail</td><td colspan="2"></td></tr>
<tr><td colspan="2">Family member and family contact</td><td colspan="6"></td></tr>
</table>

Vocabulary

flora and fauna [ˈflɔːrə ænd ˈfɔːnə]

n. 动植物；动植物群

breed [briːd]

vi. 产仔；繁殖，旺盛生长

n. 属；种类；类型；血统

livestock [ˈlaɪvstɒk]

n. 家畜，牲畜

captivity [kæpˈtɪvɪtɪ]

n. 囚禁；被关

by-products

副产品

marten [ˈmɑːtɪn]

n. 貂；貂皮

supply to

向……提供/供应……

raw material

原材料
manure [mə'njʊə]
n. 肥料；粪肥
adopt to
采取
refer to
指的是，参考，涉及
input ['ɪnpʊt]
n. 投入
speciality [speʃɪ'ælɪtɪ]
n. 专业，专长；特性
（复 specialities）
characteristic [ˌkærəktə'rɪstɪk]
adj. 特有的；独特的；表示特性的
n. 特性，特征，特色
transition [træn'zɪʃn]
n. 过渡；转变
owe to
归功于，由于
variation [veərɪ'eɪʃ(ə)n]
n. 变化；[生物] 变异，变种
accommodate [ə'kɒmədeɪt]
vt. 容纳；使适应
industrialization [ɪnˌdʌstrɪəlaɪ'zeɪʃn]
n. 工业化
behaviour [bɪ'heɪvjə]
n. 行为；举止
domesticate [də'mestɪkeɪt]
vt. 驯养；教化
thrive ['θraɪv]
vi. 兴盛，兴隆；长得健壮；茁壮成长
stable ['steɪbl]
adj. 稳定的；牢固的；坚定的
shed [ʃed]
n. 小屋，棚
slope [sləʊp]
n. 斜坡；倾斜

Unit 2

Breeds of Livestock and Poultry

Upon completing this unit, you will be able to

◆ *identify and understand the key English words or terms related to livestock and poultry breeds;*

◆ *master the English sentence structures;*

◆ *communicate in relevant activities.*

Warming up

1. What animals are these?

A. ________ B. ________ C. ________

D. ________ E. ________ F. ________

G. ________ H. ________ I. ________

2. Write the corresponding letters after the livestock or poultry according to the above pictures.

Livestock: ______________________________

Poultry: ______________________________

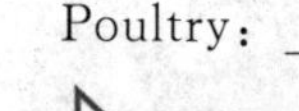

Listening and Speaking

Words and Expressions

Texel ['teksəl] sheep
n. 特克塞尔羊
White Leghorn [hwait'leghɔːn]
n. 白来杭鸡
New Hampshire ['hæmpʃiə]
n. 新罕布什尔鸡
Peking [piː'kiŋ] duck
n. 北京鸭
hatch [hætʃ]
v. 孵化；孵出

Part A　Complete the following tasks according to the instructions.

1. Listen to the dialogue and judge whether the information you hear is true (T) or false (F).

(1) He is a farm worker on the farm. (　)

(2) She is his boss. (　)

(3) There is an ox on the farm. ( )

(4) He was drawing a picture on the wall. ( )

(5) There are Texel sheep on the farm. ( )

2. Listen to the dialogue again until you can write down the sentences (W: woman, M: man).

W: ____________________

M: ____________________

W: ____________________

M: ____________________

W: ____________________

M: ____________________

W: ____________________

M: ____________________

3. Work in pairs. Ask and answer the following questions.

(1) What animals were mentioned in the dialogue?

Which one of these animals do you like most? Why?

(2) What is your occupation?

Part B Finish the following tasks.

1. Listen to the dialogue and fill in the form according to what you hear.

White Leghorns	New Hampshires	chickens in total		geese	money in total (dollars)	method of payment
			500			

2. Do role-play and practice your oral English. Suppose you are buying some livestock for your family or company.

Livestock and Poultry Breeds

A breed is a group of domestic animals that, through selection and breeding, have come to resemble one another and pass those traits uniformly on to their offspring.

In the world, almost all species of livestock and poultry are divided into different breeds. From a breeding point of view, these include heritage breeds, which are generally better suited to small-scale homesteads and modern improved ones, which are superior in the context of large-scale operation.

What Are Heritage Breeds?

There is no official universal definition or certification for "heritage" livestock and poultry, however, they are generally considered to be historic breeds, raised by our forefathers, which have been around for decades or even hundreds of years. These breeds were carefully selected and developed over time with traits that made them well-adapted to the local environment. They thrived under farming practices and cultural conditions that are very different from those found in modern agriculture.

The remarkable features of the heritage breeds are that they have a better taste but smaller bodies (e. g. Fragrance pig, Jinhua pig, Peking duck). Compared to modern cattle breeds, heritage cattle breeds often produce both decent meat and milk, and thrive on a grass-fed diet with less of a need for supplemental feed. Similarly, in comparison to today's chicken breeds that are selected for rapid growth, the heritage chicken breeds can be both prolific egg layers and solid meat birds, generally only requiring a longer time to reach full maturity.

Modern Improved Breeds

Modern improved breeds are mainly cross breeds and exotic breeds. Many breeds used in large scale agriculture have been specifically selected for intensive production including rapid growth, feed efficiency, continuous milk or egg production, or other targeted production characteristics (e. g. , the Yorkshire pig and/or Duroc are often selected to gain more lean meat percentage; for more eggs, the White Leghorn is chosen; the Holland Holstein cattle is recognized for superior milk production, producing a large quantity of milk).

In conclusion, different animal breeds are chosen depending on their intended purpose or use.

1. Write down all words you do not know in the passage. Verify the pronunciations and read them aloud.

____

____

____

2. Read the passage fluently and check your understanding by filling with proper words.

(1) ________ are generally better suited to small-scale homesteads.

(2) Fragrance pig, Jinhua pig are ________ breeds.

(3) Heritage cattle breeds can thrive on a grass-fed diet with less ________ in comparison to modern cattle breeds.

(4) Heritage breeds can adapt to local environment easier than ________ .

(5) Livestock and poultry breeds are chosen depending on ________ .

3. What livestock and poultry breeds do you know? Write down their characteristics.

____

____

____

____

Grammar

句子的种类和基本结构

1. 句子的种类

英语句子按使用目的和语气可分为：陈述句、疑问句、祈使句和感叹句。按结构可以分为：简单句（只有一套主谓结构）、并列句（包含两套或两套以上主谓结构，且没有主次之分）、复合句（包含一个主句和一个或几个从句，从句由从属连词引导）。

复合句实质上就是主句的某些成分，不是用词或短语来承担，而是用句子（从句）来承担。承担主句某些成分的句子就是从句。例如：

The film had begun *when* we got to the cinema.

主语（主句） （*关联词*）时间状语从句

2. 句子的基本结构

千变万化的英语句子归根结底都是由以下五种基本结构的简单句组合、扩展、变化而来的。

(1) 主语＋谓语（动词）。

例如：Liu Xiang runs fast.

主语 谓语

(2) 主语＋系动词＋表语。

例如：Children feel very happy.

主语 系动词 表语

(3) 主语＋谓语（动词）＋宾语。

例如：My parents like cooking.

主语 谓语 宾语

（4）主语＋谓语（动词）＋宾语＋补语。

例如：Time would prove me right.

主语 谓语 宾语 宾补

（5）主语＋谓语（动词）＋间接宾语＋直接宾语。

例如：I just sent him E-mail.

主语 谓语 间宾 直宾

不管英语句子多么简单还是复杂，归根到底不外乎以上五种结构之一。例如：

Stand up!

谓语

属于第（1）种，这是省略主语（you）的“主语＋谓语”的结构。

New methods made the work easy.

主语 谓语 宾语 宾补

属于第（4）种结构的简单句。

The weatherman said (that) there would be a strong wind in the afternoon.

主语 谓语 宾语（从句）

属于第（3）种结构，只不过宾语是用从句来承担的。

I consider it possible to work out the problem in another way.

主语 谓语 宾语 宾补 真宾语（it 只是形式宾语）

仍然是第（4）种类型的句子。

Exercises

1. Determine what structures the following sentences are.

（1）They raised 10,000 geese on the farm.

______

（2）China is a developing country; America is a developed country.

______

（3）He offered me his seat.

______

（4）In order to catch up with the others, I must work harder.

______

（5）You must work hard if you are afraid of failing.

______

2. Analyze the underlined sentences in the following passage and then translate the paragraph into Chinese.

The pig was one of the earliest animals to be domesticated in China. Pig breeding has a long history in China and the Chinese art of breeding was so skillful that numerous pig breeds of high quality were produced. China has provided the world with a fair number of the most cherished breeds of pigs. The improved pig breeds of ancient Rome, Britain and the USA, which have had a tremendous influence on pig breeding in nearly every part of the world, were themselves influenced by breeding stocks imported from southern China. The numerous breeds and varieties of pigs in China furnish a wealth of genetic material for further improvement of the livestock in our own country and others.

____

____

____

____

3. Translate the following sentences into English.

(1) 来杭鸡产蛋量很高。

____

(2) 他们要送我们 100 只绵羊。

____

(3) 我们农场有 90 头荷斯坦牛。

____

(4) 杜洛克是高瘦肉率的猪。

____

(5) 人们认为汉普夏猪是很好的瘦肉品种。

____

Extra Reading

There are two passages followed five questions, respectively. Read them and then circle the correct answer A, B, C or D for questions 1-10.

The cattle industry represents a major sector of the agricultural economy and is a large part of the American economy. For numerous years, farmers have raised cows on their farms as a family business. Even today there are still some small farms that raise cattle mostly for meat and milk, as well as for leather, household pets and other purposes.

These family farms will often enter their cows to be shown off in what is known as a cow exhibition. There are some contrasting breeds of cattle that are shown off in these shows, which makes it a great experience to attend. Numerous common breeds are displayed including: Holstein, Angus, Charolais, Limousin,

Simmental, and Shorthorn cattle.

Holstein cattle are raised for milk. This breed is the best dairy cattle because it produces more milk than other breeds. Angus cattle is the most popular cattle breed. They are notable for the quality of their beef that can be savored at some of the best restaurants. Though they are slightly smaller than many other breeds, they are considered to be the best tasting. Charolais, Limousin are also high-quality beef cattle. Simmental, and Shorthorn cattle are developed as dual-purpose, both dairy and beef production.

1. Raising cattle in America is mainly for ________ .

A. cow exhibitions

B. milk

C. meat

D. meat and milk

2. There are some contrasting breeds of cattle that are shown off in these shows, which makes it a great experience to attend. "Shown off" means ________ .

A. displayed proudly

B. give off

C. get off

D. leave off

3. Holstein cattle are ________ .

A. from America

B. famous

C. raised for dairy and milk

D. the breed which produces the most milk

4. Which one of the following statements is wrong? ________

A. Shorthorn cattle are developed for the dual-purpose of both dairy and beef production.

B. Angus cattle are smaller than many other breeds, but their beef is the best tasting.

C. Charolais produces great milk.

D. Holstein cattle are raised as the best dairy cattle.

5. Which is the best title for the passage? ________ .

A. The Cattle Industry is a Great Business in America

B. Brief Introduction of Cattle Industry in America

C. Many Breeds of Cattle in the Shows

D. Raising Cows for Shows

We preserve our old tractors, so why not preserve our old breeds of livestock and poultry too? We have some very old breeds of livestock and poultry, some of which are becoming so rare that they are in serious danger of extinction. Breeds become rare, either because their characteristics do not meet present-day demands, or because their special qualities have not yet been discovered.

Gloucestershire Old Spot Pig

Now, you may be scratching your head and thinking, "but I see cows and chickens all the time, how can they be endangered?" Just as there are endangered species such as pandas, tigers, and elephants, there are also endangered livestock and poultry breeds.

Although rare breeds exist in relatively small populations, they can make a valuable contribution, both to animal husbandry and to the quality of human life.

Some have distinctive characteristics, which can have value for protecting the environment, and others have qualities which make them well suited to less intensive farming methods. Even those which have no apparent value at present may well possess characteristics that will be important in the future and they need to be preserved as insurance against changing circumstances.

The most important reason for protecting old breeds is that it can help preserve genetic resources for future generations.

6. Which one should be the best title for the passage? ________ .

A. The reasons for preserving endangered breeds of livestock and poultry

B. How to protect rare breeds of livestock and poultry

C. Rare breeds are delicious

D. The importance of changing circumstances

7. Some breeds become rare because ________ .

A. they are too old

B. their characteristics do not suit today's demands or their special qualities have not yet been discovered

C. they have no value for protecting the environment

D. they can't make a valuable contribution, both to animal husbandry and to the quality of human life

8. "But I see cows and chickens all the time, how can they be endangered?" means ________ .

A. you can't believe the fact that they are endangered

B. you can see cows and chicken all the time, so they won't become endangered

C. you can't let them become extinct

D. you see cows and chickens become endangered

9. Some old breeds have distinctive characteristics, which make them well suited to less intensive farming methods. "Distinctive" means ________ .

A. common

B. interesting

C. of a feature that helps to distinguish a person or thing

D. very beautiful

10. The first important reason for protecting old breeds is ________ .

A. protecting the environment

B. that they can make a valuable contribution to quality of life

C. that make them well suited to less intensive farming methods

D. helping preserve genetic resources for future generations

Writing Practice

You plan to buy some livestock for your farm. But you are not sure if you can get the exact breeds you need from the company. You need to write an E-mail to the Marketing Manager Mr. Henry before you go there.

____

____

____

____

Vocabulary

poultry [ˈpəʊltri]
n. 〈集合词〉家禽

domestic [dəˈmestɪk]
adj. 家庭的，家的；国内的；驯养的

resemble [rɪˈzembl]
vt. 与…相像，类似于

trait [treɪt]
n. 特点，特性；少许

uniformly [ˈjuːnɪfɔːmlɪ]
adv. 一致地；相同地

offspring [ˈɒfsprɪŋ]
n. 后代，子孙；产物，结果；（动物的）崽

heritage [ˈherɪtɪdʒ]
n. 遗产；继承物；传统；文化遗产

homesteads [ˈhəʊmˌstedz]
n. 家宅，宅地，农庄

superiority [suːˌpɪəriˈɒrəti]
n. 优越（性），优等；傲慢

certification [ˌsɜːtɪfɪˈkeɪʃn]
n. 证明，鉴定，证书

forefather [ˈfɔː fɑːðə(r)]
n. 祖先，祖宗

decade [ˈdekeɪd]
n. 十年，十年间；十个一组；十年期

remarkable [rɪˈmɑːkəbl]
adj. 异常的，引人注目的；卓越的；非常（好）的

Fragrance pig [ˈfreɪgrəns pig]
n. 香猪

decent [ˈdiːsnt]
adj. 正派的；得体的

supplemental [ˌsʌplɪˈmentl]
adj. 补足的，追加的

prolific [prəlɪfɪk]
adj. 富饶的；（植物、动物等）丰硕的
maturity [mə'tʃʊərəti]
n. 成熟；完备
intensive [ɪn'tensɪv]
adj. 加强的，强烈的；[农] 精耕细作的；（农业方法）集约的
targeted ['tɑːgɪtɪd]
adj. 定向的
Yorkshire pig ['jɔːkʃə pig]
n. 约克夏猪
Duroc ['djʊərɔk]
n. 杜洛克猪
Holland ['hɒlənd]
n. 荷兰

Unit 3

Basic Knowledge of Anatomy

Upon completing this unit, you will be able to

◆ *identify and understand the key English words or terms related to the anatomy of livestock and poultry;*

◆ *master English sentence constituents;*

◆ *describe main structures of livestock and poultry;*

◆ *communicate in relevant activities.*

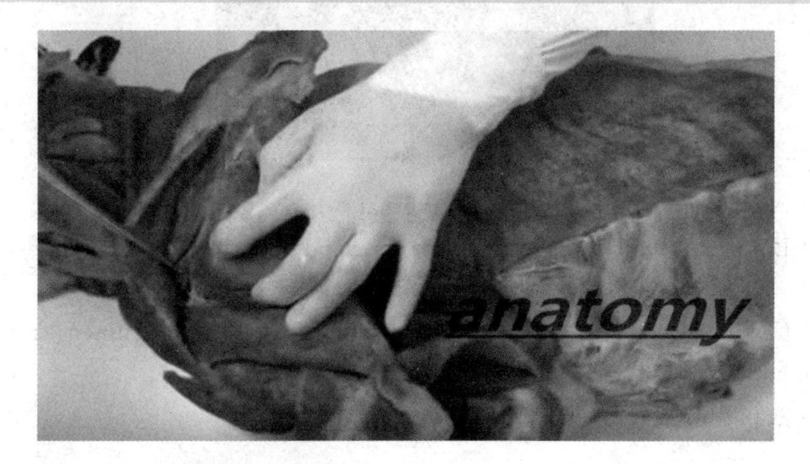

Warming up

1. Match the pictures with the correct words.

A horse
B goat
C pig
D chicken
E cattle

2. Do you think the following statements are true (T) or false (F) according to the pictures and the anatomy knowledge you have?

(1) A horse has big cecum. ( )

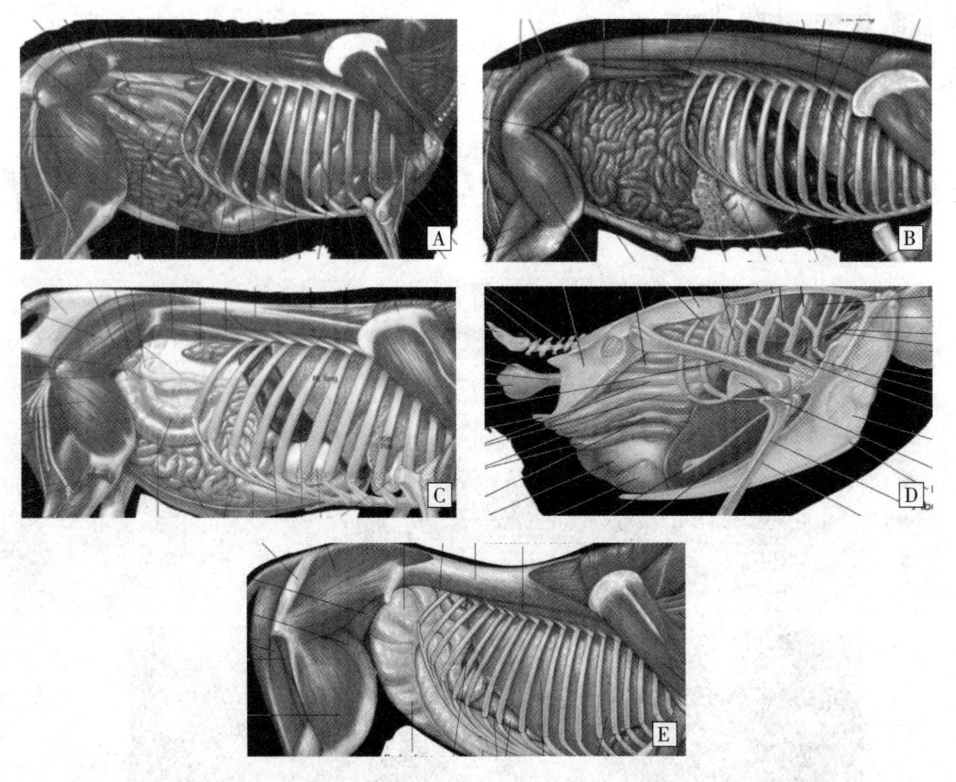

(2) A chicken's crop is the first internal organ. ( )

(3) Both cattle and goat are ruminants, and they have similar digestive systems. ( )

(4) Cattle, sheep and horse all have 4 stomachs, they are rumen, reticulum, omasum and abomasum. ( )

(5) A pig has a long small intestine. ( )

Listening and Speaking

Words and Expressions

anatomize [ə'nætəmaɪz]
v. 解剖；仔细分析
ruminant ['ruːmɪnənt]
n. 反刍动物
herbivore ['hɜːbɪvɔː(r)]
n. 食草动物
omnivore ['ɒmnɪvɔː(r)]
n. 杂食动物
carnivore ['kɑːnɪvɔː(r)]
n. 食肉动物

Part A Complete the following tasks according to the instructions.

1. Listen to the dialogue and judge whether the information you hear are true (T) or false (F).

(1) Lisa should be a student. ( )

(2) Henry is Lisa's teacher. ( )

(3) Lisa and Henry are classmates. ( )

(4) Goats, pigs and cattle were mentioned in the dialogue. ( )

(5) Lisa saw the stomach with 4 compartments when the cow was anatomized. ( )

2. Listen to the dialogue again until you can write the missing words in the blanks.

(1) Why do you look __________?

(2) I am __________ the class I attended this morning.

(3) It is an anatomy class. At the class, we __________ one pig and one goat.

(4) They both __________ of hair, skin, bone, muscle and body fluid.

(5) There is a thing that __________ me. The goat had quite a big stomach with 4 compartments, although it only had a small body.

(6) Tomorrow I will also have an __________ class.

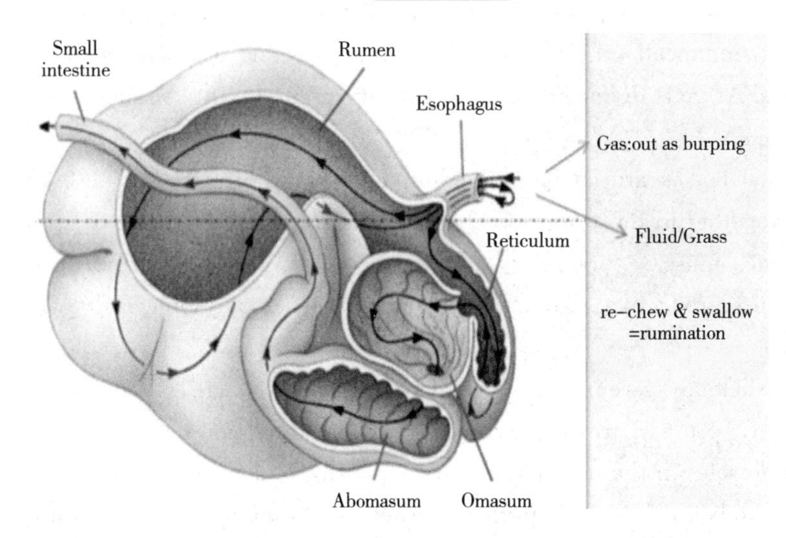

3. Work in pairs. Ask and answer questions according to the picture.

Part B Complete the following tasks.

1. Listen to the dialogue and fill in the form with Y (yes) and N (no) according to what you hear and the knowledge you have.

Types of animals	Ruminants	Non-ruminants	Herbivores	Carnivores	Omnivores
Swine	N			N	
Dogs		Y			N
Sheep	Y		Y		
Goat		N		N	
Cats			N		N
Cattle	Y			N	
Horses	N				N
Poultry		Y		N	

2. Do role-play and practice your oral English. Suppose you are participating in an anatomy class and attempting to find out the structure of the animal that is being dissected by asking your teacher.

A Brief Overview of Livestock and Poultry Anatomy

Livestock and poultry, like all higher organisms, are made of cells. They begin as a single cell (the fertilized egg or ovum) and develop into multicellular organisms. As cells divide and grow, they differentiate into tissues with a variety of functions.

Animal bodies are composed of different kinds of tissues. These tissues are grouped together to form organs. These organs are organized into systems, such as the skeletal, muscle, circulatory, respiratory, nervous, endocrine, urinary, digestive and reproductive systems. Each system carries out some major functions in the body.

The skeletal system is composed of bones, cartilage, teeth, and joints. It gives form, protection, support, and strength to the body. There are three kinds of muscles in the body: skeletal, smooth, and cardiac. The muscle system helps the body to move and perform other vital functions, such as maintaining the beating of the heart. The heart is a part of the circulatory system. The heart, arteries, capillaries, and veins make up the major parts of the circulatory system. The heart acts as a pump to keep blood moving throughout the body. Blood carries oxygen and nutrients to the cells where they are needed, and removes waste products from those cells. The respiratory system draws air into the body, where the oxygen is extracted to oxidize molecules to provide energy for the body.

Air carrying waste products, such as carbon dioxide, is then expelled out of the body. The main respiratory organs are: naval cavity, trachea, bronchi, and lungs. The nervous system provides a method for cells to transmit signals from one part of the body to another as needed. The two major parts of the nervous system are the central nervous system and the peripheral nervous system. The endocrine system secretes hormones that are needed for growth and the development of the body. The urinary system carries some waste products out of the body. The major components of the urinary system are the kidneys, ureter, bladder, and urethra. The digestive system (or tract) consists of the parts of the body involved in chewing and digesting food. Namely: the mouth, esophagus, stomach, small intestine, large intestine, rectum, and anus. The capacities of digestive systems vary greatly among different species of animals because of the different structures of digestive organs. The reproductive system refers to the sex organs of an organism which facilitate the creation of offspring.

The survival of the organism depends on the integrated activity of all the organ systems, often coordinated by the endocrine and nervous systems.

Your tasks

1. Write down all words you do not know in the passage. Verify the pronunciations of all the words and read them aloud.

____

____

2. Read the passage fluently and check your understanding by judging these statements to be true (T) or false (F). Explain your rationale.

(1) Livestock and poultry are like all higher organisms. ( )

(2) The organs of poultry are composed of systems. ( )

(3) The heart is a part of the circulatory system and acts as a pump to keep blood moving throughout the body. ( )

(4) The digestive tract consists of the mouth, esophagus, stomach, small intestine, large intestine, rectum, and anus. ( )

(5) The endocrine and nervous systems coordinate all the other organs' activities. ( )

3. Why do we study the anatomy and physiology of livestock and poultry?

____

____

词与句子成分的关系

要判断一个句子是否正确，首先要看它是否符合五种基本句型的其中一种。然后看该句的各个句子成分的用词是正确。例如：Me likes write poems. 我们可以判断它是“主-谓-宾”结构，但是该句的主语、谓语和宾语的用词都有误。正确的结构是：I like writing poems. 因为主语不能用代词宾格 me，应该用代词主格 I；谓语动词 like 不能加 s；宾语不能用动词词组 write poems，而应该把动词 write 变成动名词 writing。

试分析下列句子：

The students there are from Beijing or Shanghai.

主语（*名*）　↓　系（*动*）　表语（*介词短语*）

定语（*副*）

An animal's body is composed of a number of different kinds of tissues.

↓　主语（*名*）　谓语（*动词的被动语态*）　↓　宾语（*名*）

定语（*名词所有格*）　定语（*形容词性词组*）

The muscle system helps the body to move and perform other vital functions.

定语（*名*）↓　谓语（*动*）↓　宾语补足语（*不定式*）

主语（*名*）　宾语（*名*）

熟记英语词类与句子成分的关系对照表，有助于我们判断英语句子是否正确。

词类			句子成分						
			主语 *	谓语	宾语 *	表语 *	定语 *	状语 *	宾语/主语补足语
实词	名词		√	×	√	√	√	×	√
	代词		√	×	√	√	√	×	×
	形容词		×	×	×	√	√	×	√
	数词		√	×	√	√	√	×	√
	动词	时态语态形式	×	√	×	×	×	×	×
		动词不定式	√	×	√	√	√	√	√
		动名词	√	×	√	√	√	×	×
		动词现在分词	×	×	×	√	√	√	√
		动词过去分词	×	×	×	√	√	√	√
	副词		×	×	×	√	√	√	√

（续）

词类		句子成分						
		主语 *	谓语	宾语 *	表语 *	定语 *	状语 *	宾语/主语 补足语
虚词	感叹词/冠词 /连词/介词	×	×	×	×	×	×	×

注：“√”表示某种实/虚词可以充当句子的某个成分。

“×”表示某种实词不能充当句子的某个成分，或者某些虚词（介词、连词和冠词）不能单独充当句子的成分。感叹词只能作为独立成分，不在句子中担任成分。

“ * ”表示该成分可以由从句充当，如主语从句、定语从句、表语从句等。

Exercises

1. Correct the following sentences and explain the changes.

(1) The fruits are small in sizes, but juicy and taste.

(2) In the past we bred wonderfully breeds of cows, and many still exist today.

(3) Now people can get a lot of informations from the Internet.

(4) When a rabbit see something dangerous, it runs away.

(5) It will cost us many years to achieve a good breed of cows.

2. Analyze the use of the underlined words or expressions in the following passage and then translate the paragraph into Chinese.

The reproductive system of the female chicken is in two parts: the ovary and oviduct. Unlike most female animals, which have two functioning ovaries, the chicken usually has only one. The right ovary stops developing when the female chick hatches, but the left one continues to mature.

The ovary is a cluster of sacs attached to the hens back about midway between the neck and the tail. The oviduct is a tube like an organ lying along the backbone between the ovary and the tail.

3. Translate the following sentences into English.

(1) 动物解剖知识对于我们科学地饲养动物是至关重要的。

___

(2) 在今天的实验课上，我学到了许多有关解剖的重要细节。

___

(3) 消化系统（消化道）由用来咀嚼和消化食物的身体部位组成。

___

(4) 禽类生殖系统是不同于哺乳动物的。

___

(5) 瘤胃臌气常发生于牛、羊等草食动物。

___

Extra Reading

There are two passages followed by five questions respectively. Read them and then circle the correct answer A, B, C or D for questions 1-10.

Why can birds fly when they flap their wings, but no matter how much we flap our arms, nothing happens?

Birds can fly because their wings create airfoils that split the air. Human arms, however, are not well-shaped for creating airfoils. As we've observed, even if a strong man tried to flap his arms, he would not fly. If we look closely at a bird's wing from the side, we'll notice that it curves, if we follow along its edge, we can see that it is larger on one end than on the other. The shape of the whole wing, along with some of the individual feathers, helps make it possible for the bird to fly.

The physiological structures of birds' bodies show many unique adaptations which aid flight. Birds have a light skeletal system and powerful musculature which, along with circulatory and respiratory systems, produce high metabolic rates and oxygen supply, and thus, permit them to fly.

The bird skeleton is highly adapted for flight. It is extremely lightweight but strong enough to withstand the stresses of taking off, flying, and landing. One key adaptation is the fusing of bones into single ossification such as the pygostyle. Because of this, birds usually have a smaller number of bones than other terrestrial vertebrates. Birds also lack teeth or even a true jaw, instead they possess a beak, which is far lighter. The beaks of many baby birds have a projection called an egg tooth which facilitates their exit from the amniotic egg, and which falls off once it has done its job.

Birds have many bones that are hollow pneumatized with crisis-crossing struts or trusses for structural strength. The number of hollow bones varies among species, though large gliding and soaring birds tend to have the most. Respiratory air sacs often form air pockets within the semi-hollow bones of the bird's skeleton. All of these enable birds to fly.

Although humans can't fly, they can still do all kinds of amazing things: thinking, inventing, and using their skillful fingers to make airplanes that can help us travel, explore, and get a bird's eye view of the world.

1. Which title is the best for the passage? ________

A. The reasons why humans can't fly.

B. The reasons why birds can fly.

C. The differences between human and birds.

D. The physiological structure of birds.

2. What make it possible for the bird to fly? ________

A. The individual feathers.

B. Birds' wings create airfoils that split the air.

C. A bird's wing is larger on one end than the other.

D. All of the above.

3. What physiological structure permits birds to fly? ________

A. A light skeletal system and powerful musculature.

B. The circulatory and respiratory systems which produce high metabolic rates and oxygen supply.

C. Both A and B.

D. None of the above.

4. Which statement is correct? ________

A. The bird skeleton is light but strong.

B. Birds lack teeth but they have a true jaw.

C. Birds have greater bones than other terrestrial vertebrates.

D. Birds can fly because they have pygostyle.

5. Humans can't fly because they ________.

A. do not have feathers

B. do not have wings

C. do not have right physiological structures for flight

D. need to use their hands

Humans and some animal share the same basic muscles and bones, but they appear at different sizes, proportion and ratios based on the animal.

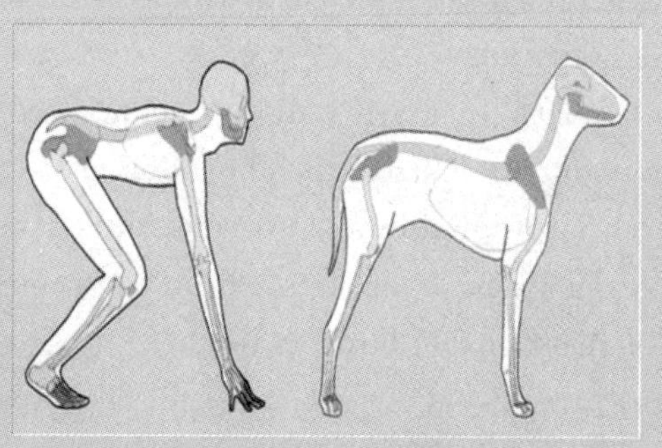

Humans and dogs have the same groups of bones and muscles even though they have completely different forms of locomotion. Using the same parts of two animals as a comparison, human's hands and dog's paws, when seen side by side, share the exact same bones in different places.

For example, pigs don't look like humans, but some researchers say pigs may someday save human lives because their organs and ours are very much alike. A comparison of internal organs, the entire inside of a pig is very similar to a human's. The hearts are in the same position between the lungs, the liver and gallbladder bear an incredible resemblance to human beings', the whole digestive systems are extremely analogous. The urinary system is also nearly identical. So, pig hearts are used to study the anatomy of human hearts because of similarities in structure, size and function. These, combined with the fact that they are much more readily available than human hearts, make them an ideal choice for research and study.

However, there are still key obstacles to replace human organs with those of another species. The immune response, or the rejection of a pig organ, is much stronger than the rejection of a human organ. Another concern is that a pig virus might infect humans. There are still big risks, unknown viruses, for example, that might be transmitted to humans. As a result, here exist questions and concerns about the use of pigs as organ donors. It will be years before pig hearts are ready for human patients.

6. What is the subject of this passage about? ________

A. Someday, pigs may save human lives because their organs and ours are similar.

B. Dogs have the same bone and muscle as humans do.

C. Pigs have saved human lives.

D. Pigs and human beings share the same internal organs.

7. They have completely different forms of locomotion. In this sentence, the underlined word "locomotion" means: ________ .

A. location

B. motion

C. revolution

D. vocation

8. Which statement is proper according to the passage? ________

A. Pigs do not like human beings.

B. Dogs like human beings.

C. A human's heart will replace the one of a pig easily.

D. Pigs and dogs have similar organs to humans.

9. Pig hearts are used to study the anatomy of human hearts because ________ .

A. pig hearts and the ones of humans are similar in structure, size and function

B. pig hearts are smaller than human ones

C. they are much more readily available than human hearts

D. Both A and C

10. The major barrier for us to use pig organs to replace those of humans are: ________ .

A. strong immune response

B. that pig virus might infect human being

C. little pig organ

D. Both A and B

Writing Practice

You are a student of a vocational college, you have been helping your teacher with making specimen of animal organs, you are asked to write the introductions of the below organs' functions.

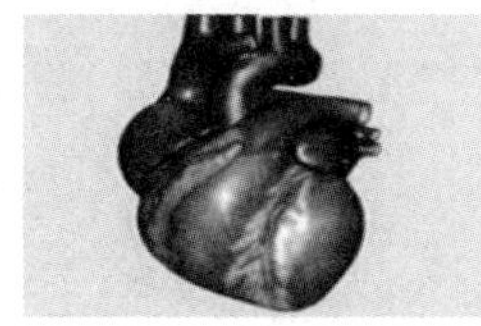
heart

stomach

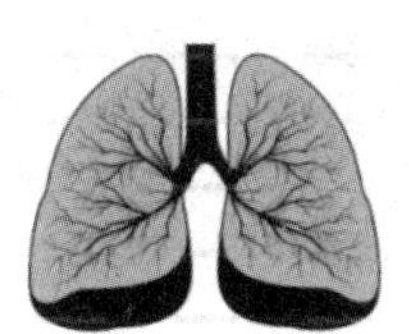
lung

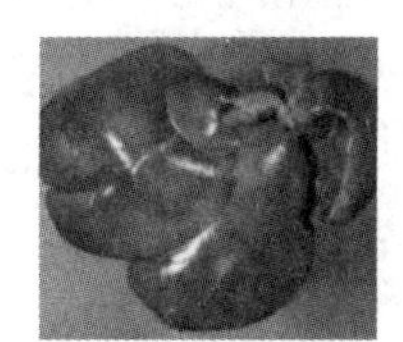
liver

____

____

Vocabulary

organism [ˈɔːgənɪzəm]
n. 有机体；生物体；微生物
fertilize [ˈfɜːtəlaɪz]
vt. 使肥沃；使受孕；施肥
ovum [ˈəʊvəm]
n. 卵子，卵细胞
multicellular [mʌltɪˈseljʊlə]
adj. 多细胞的
differentiate [ˌdɪfəˈrenʃieɪt]
vt. & vi. 区分，区别，辨别，分化
tissue [ˈtɪʃuː]
n. 薄纸，棉纸；[生] 组织；一套
skeletal [ˈskelətl]
adj. 骨骼的，骸骨的；骨瘦如柴的
circulatory [ˈsɜːkjəˌleɪtəri]
adj. （血液或汁液）循环的
respiratory [rəˈspɪrətri]
adj. 呼吸的
endocrine [ˈendəʊkrɪn]
adj. 内分泌（腺）的，激素的
urinary [ˈjʊərɪnəri]
adj. 尿的；泌尿器的；泌尿的
reproductive [ˌriːprəˈdʌktɪv]
adj. 生殖的；再生产的；复制的
cartilage [ˈkɑːtɪlɪdʒ]
n. [解] 软骨；软骨结构
joint [dʒɔɪnt]
n. [解] 关节；接合处；下流场所
cardiac [ˈkɑːdiæk]
adj. 心脏（病）的；（胃的）贲门的
oxidize [ˈɒksɪdaɪz]
vt. 使氧化；使生锈；vi. 氧化；变成氧化的
molecule [ˈmɒlɪkjuːl]
n. 分子；微小颗粒
carbon dioxide [ˈkɑːbən daɪˈɔksaɪd]
n. 二氧化碳
naval cavity [ˈneɪvl kævəti]
n. 鼻腔
trachea [trəˈkiːə]
n. 气管；导管；螺旋纹管
bronchi [ˈbrɒŋkaɪ]
n. （bronchus 的名词复数）支气管
peripheral [pəˈrɪfərəl]
adj. 外围的；（神经）末梢区域的
secrete [sɪˈkriːt]
vt. [生] 分泌；隐匿，隐藏；私下侵吞
hormone [ˈhɔːməʊn]
n. 激素
kidney [ˈkɪdni]
n. 肾；性格
ureter [jʊˈriːtə]
n. 尿管，输尿管
bladder [ˈblædə(r)]
n. 膀胱；囊状物
urethra [jʊˈriːθrə]
n. 尿道
esophagus [ɪˈsɒfəgəs]
n. 食管，食道

intestine [ɪn'testɪn]
n. [解] 肠；adj. 内部的；国内的

artery ['ɑːtəri]
n. [解剖] 动脉；干线，要道

capillary [kə'pɪləri]
n. 毛细管；毛细血管；微管

vein [veɪn]
n. 静脉；[植] 叶脉；气质，倾向

rectum ['rektəm]
n. 直肠

anus ['eɪnəs]
n. 肛门

Unit 4

Genetics and Breeding

Upon completing this unit, you will be able to

◆ *identify the English words or terms related to genetics and breeding;*

◆ *master compound sentence and complex sentence;*

◆ *communicate in relevant activities.*

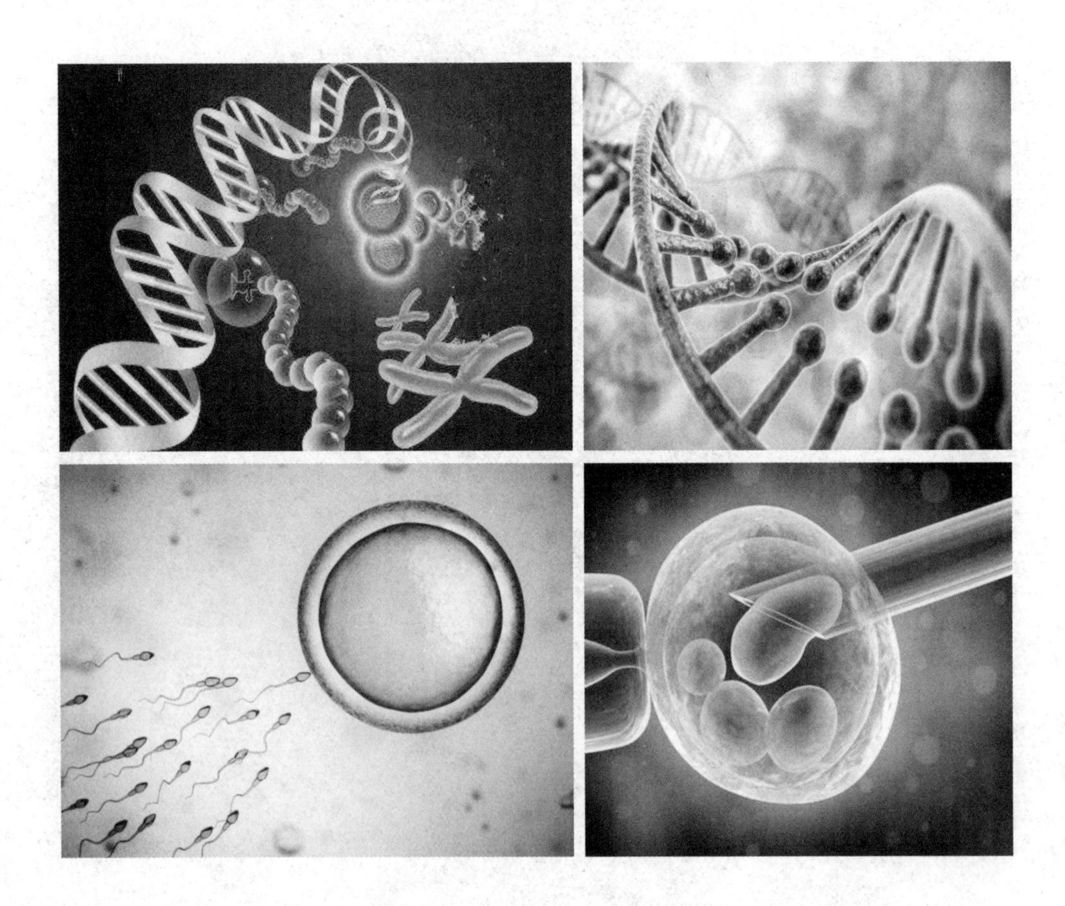

Warming up

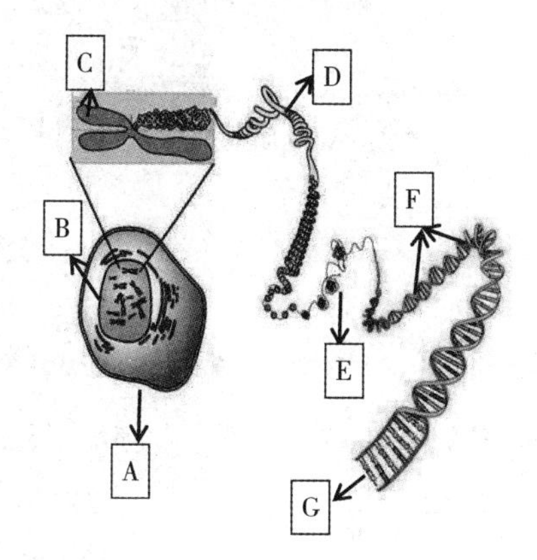

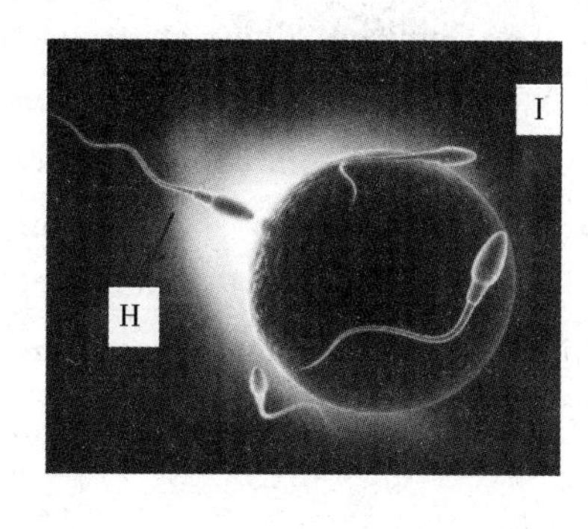

1. Do you know the above pictures? Please try to write down the corresponding words below.

A. ______________ B. ______________ C. ______________

D. ______________ E. ______________ F. ______________

G. ______________ H. ______________ I. ______________

2. What is happening in the second picture? Please rearrange the following sentences according to the process of fertilization.

(1) Sperm usually survives up to several days in uterus waiting for ovulation.

(2) Then sperm goes into the fallopian tube.

(3) Sperm and egg meet in the outer portion of the fallopian tube and the sperm enters the egg (fertilization).

(4) Sperm moves up through the cervix into the uterus.

(5) Sperm is deposited in vagina.

Listening and Speaking

Words and Expressions

slaughter [ˈslɔːtə(r)]
n. 屠宰（动物）

live-weight [liv weit]
n. 活重

carcass [ˈkɑːkəs]
n.（家畜屠宰后的）躯体

semi-eviscerated [ˈsemi ɪˈvɪsəˌreɪtid]
adj. 半切除的

abdominal [æbˈdɒmɪnl]
adj. 腹部的

Part A Complete the following tasks according to the instructions.

1. Listen to the dialogue and judge whether the information you hear are true (T) or false (F).

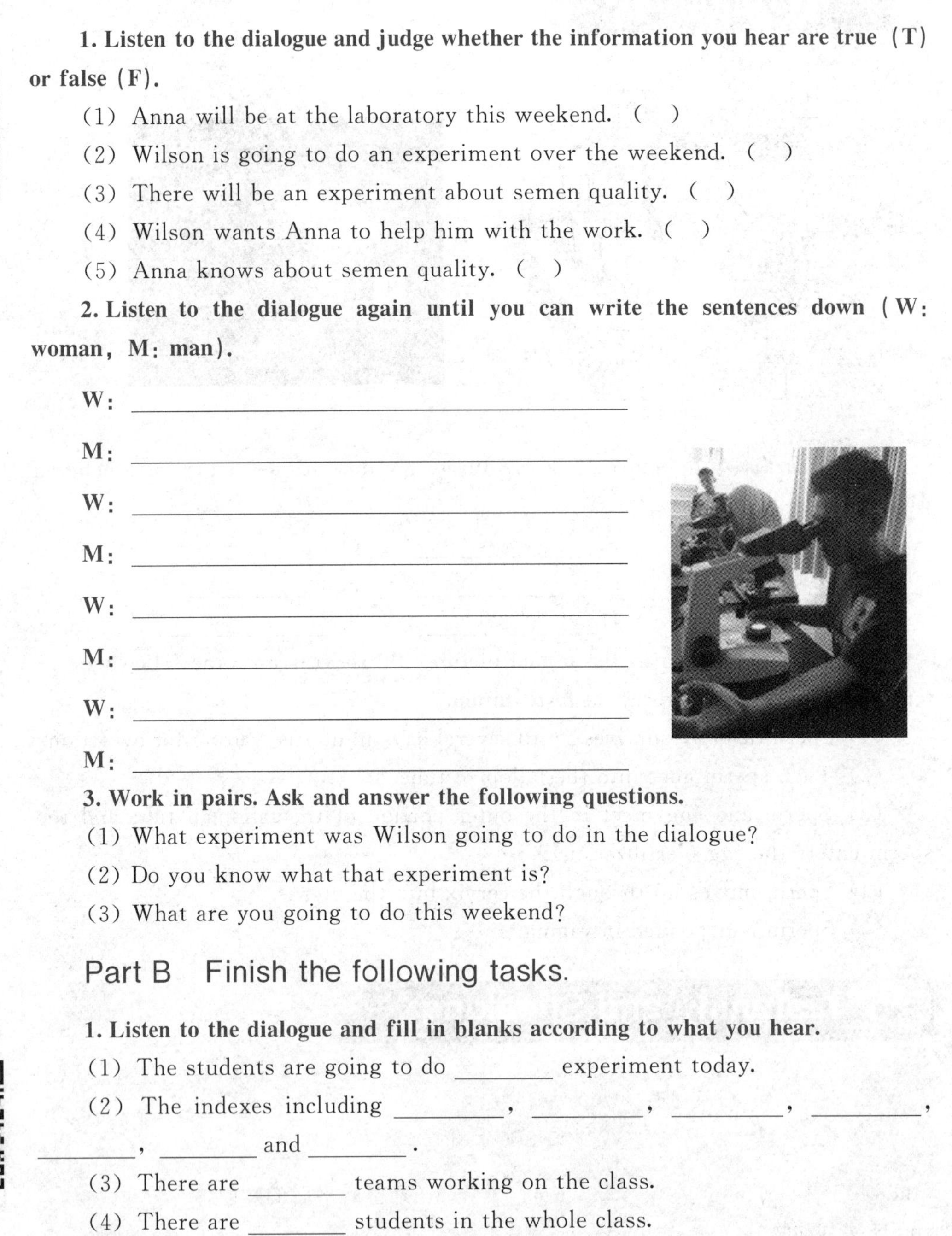

(1) Anna will be at the laboratory this weekend. ( )

(2) Wilson is going to do an experiment over the weekend. ( )

(3) There will be an experiment about semen quality. ( )

(4) Wilson wants Anna to help him with the work. ( )

(5) Anna knows about semen quality. ( )

2. Listen to the dialogue again until you can write the sentences down (W: woman, M: man).

W: ________________________________

M: ________________________________

W: ________________________________

M: ________________________________

W: ________________________________

M: ________________________________

W: ________________________________

M: ________________________________

3. Work in pairs. Ask and answer the following questions.

(1) What experiment was Wilson going to do in the dialogue?

(2) Do you know what that experiment is?

(3) What are you going to do this weekend?

Part B Finish the following tasks.

1. Listen to the dialogue and fill in blanks according to what you hear.

(1) The students are going to do ________ experiment today.

(2) The indexes including ________, ________, ________, ________, ________, ________ and ________.

(3) There are ________ teams working on the class.

(4) There are ________ students in the whole class.

2. Do role-play and practice your oral English. Talk about one of the animal husbandry and veterinary science experiments you have been learning about.

Genotype and Phenotype

Understanding Genotype and Phenotype

Wilhelm Johannsen was a scientist working in Denmark in the late 19th and early 20th centuries. During a series of experiments, he observed variations in genetically identical beans. He concluded the variation must be due to environmental factors and coined the terms "genotype" and "phenotype" in 1911.

Genotype is the genetic make-up of an individual organism. Your genotype functions as a set of instructions for the growth and development of your body. The word "genotype" is usually used when talking about the genetics of a particular trait (like eye color).

Phenotype is the observable physical or biochemical characteristics of an individual organism, determined by both genetic make-up and environmental influences, for example, height, weight and skin color.

How Genotype affects Phenotype

The term "genotype" is usually used to refer to specific alleles. Alleles are alternative forms of the same gene that occupy the same location on a chromosome. At any given locus, there are two alleles (one on each chromosome in the pair) —you get one allele from your mother and the other from your father. The two alleles might be the same or they might be different. Different alleles of a gene generally serve the same function (for example, they code for a protein that affects eye color) but may produce different phenotypes (for example, blue eyes or brown eyes) depending on which set of the two alleles one possesses.

For example, the ability to taste PTC (a bitter tasting compound) is controlled by a single gene. This gene has at least seven alleles but only two of these are commonly found.

An upper case "T" represents the dominant allele that confers the ability to taste— "dominant" means that anyone with one or two copies of this allele will be able to taste PTC. The non-tasting allele is recessive and is represented by a lower case "t" — "recessive" means that an individual will need two copies of the allele to be a non-taster.

Each pair of alleles represents the genotype of a specific individual, and in this case, there are three possible genotypes: TT (taster), Tt (taster) and tt (non-taster). If the alleles are the same (TT or tt), the genotype is homozygous. If the alleles are different (Tt), the genotype is heterozygous.

In conclusion, your genotype or genetic make-up plays a critical role in your development. However, environmental factors influence our phenotypes throughout our lives, and it is this on-going interplay between genetics and environment that makes us all unique.

Your tasks

1. Write down all words you do not know in the passage. Check the pronunciations of all the words, read them aloud.

2. Read the passage fluently and check your understanding by filling with proper words.

(1) In the late 19th and early 20th centuries, variations in ________ identical beans were studied by Wilhelm Johannsen.

(2) Wilhelm Johannsen concluded the ________ must be due to environmental factors.

(3) The genetics of a particular trait: ________.

(4) The observable physical or biochemical characteristics of an individual organism: ________.

(5) The on-going ________ between genetics and environment makes everything unique.

3. For a species of your choice, list one of its genotype and phenotype according to the passage.

Grammar

并列句、复合句

1. 并列句

并列句（Compound Sentences）指包含两个或两个以上主谓结构的句子，句与句之间没有主次之分，通常用并列连词、分号或者逗号来连接。例如：The food was good，but he had little appetite.（译文：食物很精美，但他却没什么胃口。）并列句分为：联合并列句、转折并列句、选择并列句和因果并列句。

2. 复合句

复合句（Complex Sentence）指包含两个或两个以上主谓结构的句子，句与句之间有主次之分，其中一套主谓结构为主句，其余为从句，从句通常用关系代词或关系副词与主句连接。主句是全句的主体，可以独立存在；从句则是一个句子成分，不能独立存在。从句实际上就是除谓语之外的句子成分用句子来承担，而不是用词或短语来承担。从句主要有：主语从句、表语从句、宾语从句、定语从句、状语从句。

（1）主语从句，即在复合句中充当主语成分的句子。

例句：What he did made us feel shy.

（2）表语从句就是用一个句子作为表语。

例句：What the police want to know is when you entered the room.

（3）宾语从句是在主从复合句中充当宾语，位于及物动词、介词或复合谓语之后的从句。

例句：Everybody knows that money doesn't grow on trees.

（4）定语从句是修饰或限定名词或代词的从句，被修饰的名词或代词称为先行词。定语从句分为限定性定语从句和非限定性定语从句。

例句：A prosperity which / that had never been seen before appears in the countryside.

（5）状语从句可以修饰谓语、非谓语动词、定语、状语或整个句子。状语从句主要分为时间、地点、原因、结果、方式、比较、目的、条件和让步状语从句。

例句：You will certainly succeed so long as you keep on trying.

Exercises

1. What types are the following sentences? Are they simple, compound or complex sentences?

（1）Mr. Hu is not only a businessman but also a teacher.

（2）China is a developing country; America is a developed country.

（3）Changsha is the place where I was born.

（4）We should do everything that is useful for the people.

（5）You must work hard if you are afraid of failing.

2. Analyze the underlined sentences in the following passage and then translate the paragraph into Chinese.

There are many different definitions of a breed. But they have several collective characteristics: A breed is a population of (related) animals that look alike in apparent characteristics and that also pass these apparent characteristics on to their descendants. The animals of one breed differ from animals of another breed.

Breeds are often organized in a genealogical register or breed association. They have defined several conditions (often appearance characteristics) that an animal must comply with. These conditions are called the breed standard.

3. Translate the following sentences into English.

(1) 性染色体是指携带性别遗传基因的染色体。

(2) 基因是具有遗传效应的 DNA 片段。

(3) 荷斯坦牛因产奶量高，而闻名全世界。

(4) 娟珊牛正是 Mr. White 想引进的奶牛品种。

(5) 我以为这头猪已经打过疫苗了。

There are two passages followed by five questions, respectively. Read them and then circle the correct answer A, B, C or D for questions 1-10.

There are all kinds of animals on our planet. Some are amiable, some are not; some are small, some are tall. Here are four pictures of animals with some of the funniest, friendliest faces. But remember, although they look nice, that doesn't mean they are always friendly.

Not all dogs are cute, but most of them are pretty friendly. When you come near an unfamiliar dog, remember to hold out your hand, so he/she can smell you and see that you mean no harm.

Cows love to eat grass, and they eat quite a lot! You need not be afraid of this quiet farm animal. But do not run toward at a cow in a field or touch one, unless you are told by an adult that it is okay. Horses are big, beautiful animals, but they can be surprised easily. A horse might jump up if it hears a loud noise, or kick its legs if you approach too suddenly. Different horses, like different humans, are different.

Chimpanzees make amusing faces and put on a real show at the zoo, but believe it or not, they can be very dangerous. And they are strong too, much stronger than humans—so you should never make a chimp mad!

1. Which of the following animals can make faces? ________

A. Cows.

B. Chimpanzees.

C. Dogs.

D. Horses.

2. Which one is the proper way to communicate with unfamiliar dogs? ________

A. Let the dog smell you.

B. Touch the dog.

C. Smile to the dog.

D. Stretch your hand to the dog.

3. We can learn from the passage that ________.

A. cows are quiet

B. all dogs are cute

C. horses never kick

D. chimpanzees are not dangerous

4. According to the passage, cows are a kind of animal that ________.

A. like to be touched by people

B. eat many vegetables
C. easily to get angry
D. are quiet and kind

5. The author wants to tell us that ________ .

A. some animals are sweet, but they are small
B. some animals are not sweet, but they are tall
C. some animals are funny, but they are not beautiful
D. some animals look nice, but they are not always friendly

Public opinion about zoos is divided: there are those who think it is cruel to keep animals locked up, while others believe zoos are essential for the survival of endangered species.

To argue that zoos imprison animals is to misunderstand what zoos are about. Without zoo, many of the creatures we love and admire would no longer exist. Every single day, over one hundred animal species vanish. Scientists predict that as early as 2050 one quarter of the Earth's species will become extinct.

Some animals are in danger because they are hunted. Alarmingly, the population of tigers has already fallen by ninety-five percent. Other species are in danger because of a lack of food.

Zoos have special breeding programs to help those animals at risk. These breeding programs are proving extremely successful. As Irene Shapiro from Zoo and Wildlife Support reported, "the Puerto Rican Parrot has grown in numbers from just thirteen to about eighty-five and the Golden Lion Tamarin Monkey, which has almost ceased to exist twenty years ago, has been reintroduced back into the wild."

Unfortunately, not everyone understands the important role zoos play. For example, Brian Feather stone from the Anti-Zoo Forum says, "I can't believe we take animals from the wild and put them in cages for the entertainment of the public! We should view them on film or TV in their natural environment." However, this misses the point. A zoo does more than display animals to the public, it ensures their survival. Without zoo, you would not be able to see many of these animals on TV or anywhere else!

6. What are the writer's attitude towards zoos? ________

A. He shows no opinion either for or against them.

B. He thinks that they are unnecessary and cruel.

C. He believes they play an important environmental role.

D. He expresses a desire that more of them be built.

7. The underlined word "vanish" in Paragraph 2 most probably means ________ .

A. remain

B. disappear

C. become rarer

D. get killed

8. According to the passage the world's tiger population ________ .

A. will rise by 5% next year

B. is relatively stable

C. is 95% smaller than in the past

D. has fallen to 95%

9. According to the passage some people do not agree with zoos because they ________ .

A. are too expensive to run

B. put animals in danger

C. do not provide enough food

D. keep animals locked up

10. According to the writer, the most important function of the zoo is to ________ .

A. make a lot of money

B. entertain visitors

C. ensure animals' survival

D. educate the public

Writing Practice

In your dairy farm, how can you get a cow breed with high milk production and strong hoofs? Think about the breeding method you would choose.

____

____

____

Vocabulary

genotype [ˈdʒenətaɪp]
n. 基因型；遗传型

phenotype [ˈfiːnətaɪp]
n. 显型；表现型

identical [aɪˈdentɪkl]
adj. 同一的；完全同样的，相同的

heterozygote [ˌhetərəˈzaɪgəʊt]
n. 异质接合体，异形接合体，杂合体

coin [kɔɪn]
vt. 制造硬币；杜撰；创造

individual [ˌɪndɪˈvɪdʒuəl]
adj. 单独的；一个人的；独有的

make-up [meik ʌp]
组成；补足；化妆；编造

allele [ˈæliːl]
n. 等位基因

alternative [ɔːlˈtɜːnətɪv]
adj. 替代的；备选的；其他的

chromosome [ˈkrəuməsəum]
n. [生] 染色体

locus [ˈləʊkəs]
n. 所在地，场所；核心；色点

compound [ˈkɒmpaʊnd]
n. 场地；复合物

dominant [ˈdɒmɪnənt]
adj. 占优势的；[生] 显性的

recessive [rɪˈsesɪv]
adj. （遗传特征）隐性的

homozygous [ˌhɒməˈzaɪgəs]
adj. 同型结合的，纯合子的

critical [ˈkrɪtɪkl]
adj. 批评的；关键的；极重要的

ongoing [ˈɒngəʊɪŋ]
adj. 不间断的，进行的；前进的

interplay [ˈɪntəpleɪ]
n. 相互作用

unique [juˈniːk]
adj. 唯一的，仅有的；独特的

amiable [ˈeɪmiəbl]
adj. 和蔼可亲的；温和的

chimpanzee [ˌtʃɪmpænˈziː]
n. 黑猩猩

chimp [tʃɪmp]
n.（非洲）黑猩猩

essential [ɪˈsenʃl]
adj. 基本的；必要的；本质的

extinct [ɪkˈstɪŋkt]
adj. 灭绝的；绝种的；消逝的

Unit 5

Pasture Construction

Upon completing this unit, you will be able to

- *identify the English words or terms related to pasture construction;*
- *master subject clause;*
- *communicate in relevant activities.*

Warming up

1. Write down the corresponding names according to the above pictures.

A. ____________ B. ____________ C. ____________

D. ____________ E. ____________ F. ____________

2. Choose one of the above pictures and make a brief introduction according to your own knowledge.

___

___

___

Words and Expressions

promotion [prəˈməʊʃn]
n. 促进，增进；提升，升级
ostrich [ˈɒstrɪtʃ]
n. 鸵鸟
tribal [ˈtraɪbl]
adj. 部落的；种族的
ceremonial [ˌserɪˈməʊniəl]
adj. 仪式的；正式的
cholesterol [kəˈlestərɒl]
n. 胆固醇

Part A Complete the following tasks according to the instructions.

1. Listen to the dialogue and judge whether the information you hear are true (T) or false (F).

(1) Richard has a new job now. ( )

(2) Jackson works on a farm. ( )

(3) There's a new egg production system on the farm. ( )

(4) The new egg collection system is automatic. ( )

(5) The new construction costs 400,000 yuan. ( )

2. Listen to the dialogue again until you can write the sentences down (W: woman, M: man).

W: ________________

M: ________________

W: ________________

M: ________________

W: ________________

M: ________________

W: ________________

M: ________________

W: ________________

3. Work in pairs. Ask and answer the following questions.

(1) What is the new job Jackson is doing right now?

(2) What is the new egg production construction about?

(3) Why does Jackson say it's well worth the money?

Part B Finish the following tasks.

1. Listen to the dialogue and fill in blanks according to what you hear.

(1) An ostrich ________ a wild animal.

(2) Ostrich meat has ________ protein than beef.

(3) ________ has farmed ostriches since 1860.

(4) The leather of ostrich is very ________, and has a ________ quality.

2. Do role-play and practice your oral English. Talk about what are ostriches benefits and what you can do with it.

Organic Farming

Organic farming is a method of crop and livestock production that involves much more than choosing not to use pesticides， fertilizers， genetically modified organisms， antibiotics and growth hormones.

Organic production is a holistic system designed to optimize the productivity and fitness of diverse communities within the agro-ecosystem， including soil organisms， plants， livestock and people. The principal goal of organic production is to develop enterprises that are sustainable and harmonious with the environment.

Why Farm Organically?

The main reasons farmers state for wanting to farm organically are their concerns for the environment and about working with agricultural chemicals in conventional farming systems. There is also an issue with the amount of energy used in agriculture， since many farm chemicals require energy intensive manufacturing processes that rely heavily on fossil fuels. Organic farmers find their method of farming to be profitable and personally rewarding.

Why Buy Organic?

Consumers purchase organic foods for many different reasons. Many want to buy food products that are free of chemical pesticides or grown without conventional fertilizers. Some simply like to try new and different products. Product taste， concerns for the environment and the desire to avoid foods from genetically engineered organisms are among the many other reasons some consumers prefer to buy organic food products. In 2007 it was estimated that over 60 percent of consumers bought some organic products. Approximately 5 percent of consumers are considered to be core organic consumers who buy up to 50 percent of all organic food.

Successful Organic Farming

In organic production， farmers choose not to use some of the convenient chemical tools available to other farmers. Design and management of the production system are critical to the success of the farm. Select enterprises that complement each other and choose crop rotation and tillage practices to avoid or reduce crop problems.

Yields of each organic crop vary， depending on the success of the manager. During the transition from conventional to organic， production yields are lower than conventional levels， but after a three-to-five-year transition period the organic yields typically increase.

Organic farming can be viable alternative production method for farmers, but there are many challenges. One key to success is being open to alternative organic approaches to solving production problems. Determine the cause of the problem, and assess strategies to avoid or reduce the long-term problem rather than a short-term fix for it. (引自 http://www.omafra.gov.on.ca/english/crops/facts/09-077.htm)

Your tasks

1. Write down all words you do not know in the passage. Check their pronunciations and read them aloud.

____

____

____

2. Read the passage fluently and check your understanding by filling in the blanks with the proper words.

(1) Organic farming includes ________ and ________ productions.

(2) Organic farming choose not to use ________________________.

(3) More and more farmers start organic farming, because it's ________ and ________.

(4) The principal goal of organic production is to develop enterprises that are sustainable and ________ with the environment.

(5) Design and management of the production system are ________ to the success of the farm.

3. Make a list for the benefits and shortcomings of organic farming.

____

____

____

Grammar

主 语 从 句

主语从句(subject clause)在复合句中充当主语成分，属于名词性从句。通常用陈述句、一般疑问句或特殊疑问句承担。

1. 陈述句用作主语

陈述句作主语时，引导词用 that。

例句：That the dog is still alive is a consolation.

那只狗还活着，这是令人欣慰的。

如果主语从句太长，为了句子的平衡，就用 it 作形式主语，把 that 引导的主语从句放到句末。

例句：It is a miracle that the technique of cloning a monkey succeeded in China.

克隆猴的技术在中国取得了成功，这是一个奇迹。

2. 一般疑问句作主语

一般疑问句作主语时，要把一般疑问句变成陈述句，并且用 whether 来引导。

例句：Whether he likes me or not makes no difference to me.

他喜不喜欢我，我无所谓。

3. 特殊疑问句作主语

特殊疑问句作主语时，把特殊疑问词放句首，后面是陈述句结构。

例句：Why the chickens died is still unknown to us.

我们尚不清楚鸡群死亡的原因。

这里是把原来的 Why did the chickens die? 变成了 why＋陈述句结构：the chickens died . 去掉了助动词 did，把 die 变成 died.

例句：When the sport games will be held depends on the weather.

运动会何时举行取决于天气。

这里是把原来的 When will the sport games be held? 变成了：

When＋ the sport games will be held. （主谓结构的陈述语序）

主 谓

Exercises

1. Change the following sentences into the structure of "It is... that...".

(1) That smoking can cause cancer is true.

______

(2) That he has been late for work over and over is a serious matter.

______

(3) That the world is round is a fact.

______

(4) That the seas are being overfished has been known for years.

______

(5) That that organic food is becoming more and more popular is a trend.

____

2. Fill in the blanks by choosing A, B, C or D.

(1) It is a problem ________ we do not have enough money for the experiment.

A. why B. when C. that D. whether

(2) Is this the factory ________ you visited the other day?

A. where B. when C. whether D. why

(3) ________ he said at the meeting surprised everyone present.

A. That B. What C. The matter D. Thing

(4) ________ we'll go camping tomorrow depends on the weather.

A. Why B. That C. Which D. Whether

(5) ________ is necessary that every student masters a foreign language.

A. It B. Whether C. When D. That

3. Translate the following sentences into English.

(1) 人们想要吃的是从未污染过的新鲜食物。

____

(2) 据说去年有 10 000 头母牛被引进了他们的新牧场。

____

(3) 他们是否同意我们参观农场是个问题。

____

(4) 我们该把这些猪养在哪里还没有决定。

____

(5) 家禽养殖业很赚钱是事实。

____

Extra Reading

Read the following two texts. Answer the questions on each text by choosing A, B, C or D for questions 1-10.

The production of transgenics provides methods to rapidly introduce "new" or modified genes and DNA sequences into livestock without crossbreeding or hybridizing. It is a more precise technique, but not fundamentally different from genetic selection or crossbreeding in its result. Much has been written about the methodologies used to produce transgenic livestock and that aspect will not be covered in this review. The obvious question is "Why genetically modify livestock?" The answer is not so straightforward; however, some of the reasons are to

◇study the genetic control of physiological systems;

◇build genetic disease models;

◇improve animal production traits;

◇produce new animal products.

There are many potential applications of transgenic methodology to develop new and improved strains of livestock. Practical applications of transgenics in livestock production include enhanced prolificacy and reproductive performance, increased feed utilization and growth rate, improved carcass composition, improved milk production and/or composition, modification of hair or fiber, and increased disease resistance. Development of transgenic farm animals will allow more flexibility in direct genetic manipulation of livestock. Gene transfer is a relatively rapid way of altering the genome of domestic livestock. The use of these tools will have a great impact toward improving the efficiency of livestock production and animal agriculture in a timely and more cost-effective manner. With ever-increasing world population and changing climate conditions, such effective means of increasing food production are needed. (引自: https://www.nature.com/scitable/knowledge/library/transgenic-animals-in-agriculture-105646080)

1. Which one is not the highlight of transgenic productions? ________

A. easy to do

B. rapidly introduce genes

C. precise

D. no need crossbreeding

2. "The obvious question is 'Why genetically modify livestock?' The answer is not so straightforward." Which of the following is the probable explanation for "straightforward"? ________

A. direct

B. simple

C. complex

D. unknown

3. Which one of the following statements is right? ________

A. Methodologies used to produce transgenic livestock are simple.

B. Transgenic is fundamentally different from genetic selection or crossbreeding in its result.

C. Building genetic disease models is one reason why people do transgenic.

D. Gene transfer can slowly alter the genome of domestic livestock.

4. Practical applications of transgenics in livestock production could NOT ________.

A. enhance prolificacy

B. reduce reproductive performance

C. increase feed utilization

D. improve growth rate

5. Which title is the best for the passage? ________ .

A. Transgenic livestock

B. The production of transgenics

C. How to produce transgenic animals

D. Gene transfer

What is poultry farming? Poultry farming means raising various types of domestic birds commercially for the purposes of meat, eggs and feather production. The most widely and common raised poultry is chicken. The chickens which are raised for eggs are called layer chicken; and the chickens which are raised for meat are called broiler chicken. Over 12 billion chickens are raised every year as food source in China. In a word, poultry farming is very necessary to meet the demands of animal nutrition. And the poultry farming business is also very profitable.

Free range and intensive farming are the two main methods for poultry farming. Free range farming method is used for large number of poultry birds with high stocking density. There are some basic differences between intensive and free-range poultry farming. The intensive poultry farming method is a highly efficient system which saves land, feed, labor and other resources and increases production. In this system, the poultry farming environment is fully controlled by the farmer. So, it ensures continuous production throughout the year without being affected by seasons. Intensive poultry farming has some disadvantages too. Some people say intensive system creates health risks, abuses the animals and is harmful for environment. On the other hand, free range poultry farming method requires a large area for raising the birds and the production is about the same as intensive method. However, in the case of both intensive and free-range poultry farming method the producers must regularly use nationally approved medications like antibiotics to keep the poultry birds free from diseases.

6. Which of the following is right? ________

A. There are only two methods for poultry farming.

B. Free range is better than intensive farming.

C. Chicken can be separated into two categories: layer one and broiler one.

D. By using the free range method, the chickens will not catch any diseases.

7. Which of the following is not a benefit for intensive farming? ________

A. Highly efficient.

B. Saving land.

C. Continuous production.

D. Animal welfare.

8. If you want to use less feed, you should choose ________.

A. free range

B. cage

C. intensive farming

D. all of the above

9. This text mainly discusses ________.

A. the two sides of free range poultry farming

B. comparing free range and intensive farming

C. ways for poultry farming

D. all of the above

Writing Practice

Advertisements are written to try and make us buy something. The words in an advertisement give us a very clear idea of the product, and really make us want to have it. Look at this example of an advertisement for ice cream: "Cool, delicious Strawberry Surprise. What could be nicer on a hot summer's day?"

A. Imagine that you are an advertiser who is trying to get people to buy the pork from your farm. Write the words that you would use to describe your pork (you can use your dictionary!).

1. ________ 2. ________ 3. ________ 4. ________ 5. ________

6. ________ 7. ________ 8. ________ 9. ________ 10. ________

B. Using some of your words from above, write an advert for the pork in the space below.

____

____

____

____

Vocabulary

pasture [ˈpɑː stʃə(r)]
n. 牧草地，牧场；牲畜饲养，放牧
vt. & vi. 放牧
organic [ɔːˈgænɪk]
adj. 有机（体）的；系统的；根本的
pesticide [ˈpestɪsaɪd]
n. 杀虫剂，农药
fertilizer [ˈfɜːtəlaɪzə(r)]
n. 肥料，化肥；受精媒介物
genetically modified
[dʒeˈnetikəli ˈmɔdifaid]
转基因的
antibiotics [ˌæntɪbaɪˈɒtɪks]
n. （用作复数）抗生素
holistic [həʊˈlɪstɪk]
adj. 全盘的，整体的；功能整体性的
optimize [ˈɒptɪmaɪz]
vt. 使最优化，使尽可能有效
fitness [ˈfɪtnəs]
n. 健康；适当，适合；合情理
diverse [daɪˈvɜːs]
adj. 不同的，多种多样的；变化多的
agroecosystem [ˈægrəʊ iːkəʊˈsistəm]
n. 农业生态系统
sustainable [səˈsteɪnəbl]
adj. 可持续的；可以忍受的；可支撑的
harmonious [hɑˈməʊniəs]
adj. 和谐的，融洽的；协调的
conventional [kənˈvenʃənl]
adj. 传统的；依照惯例的；约定的
manufacturing [ˌmænjuˈfæktʃərɪŋ]
n. 制造业，工业
adj. 制造业的，制造的
profitable [ˈprɒfɪtəbl]
adj. 有利可图的，有益的；合算的
consumer [kənˈsjuːmə(r)]
n. 消费者
purchase [ˈpɜːtʃəs]
n. 购买；购买行为；购置物；紧握
v. 购买；采购；换得；依靠机械力移动
estimate [ˈestɪmeɪt]
n. 估计；报价；评价，判断
vt. 估计，估算；评价；估量，估价
approximately [əˈprɒksɪmətli]
adv. 近似地，大约
core [kɔː(r)]
n. 中心，核心，精髓；果心，果核
complement [ˈkɒmplɪment]
n. 补充；补足语；补充物；补集（数）
vt. 补足，补充；补助
rotation [rəʊˈteɪʃn]
n. 旋转，转动；轮流，循环；[农]轮作
tillage [ˈtɪlɪdʒ]
n. 耕耘，耕地
viable [ˈvaɪəbl]
adj. 切实可行的；能自行生产发育的
approach [əˈprəʊtʃ]
vt. & vi. 接近，走近，靠近
n. 方法；途径；接近
assess [əˈses]
vt. 评定；估价

Unit 6

Scientific Feeding

Upon completing this unit, you will be able to

- ◆ *identify the English words or terms related to scientific raising animals;*
- ◆ *master predicative clause and object clause;*
- ◆ *communicate in relevant activities.*

Warming up

1. Do you know the feed of the rabbit? Write down them on the blank.

A. ________________ B. ________________

C. ________________ D. ________________

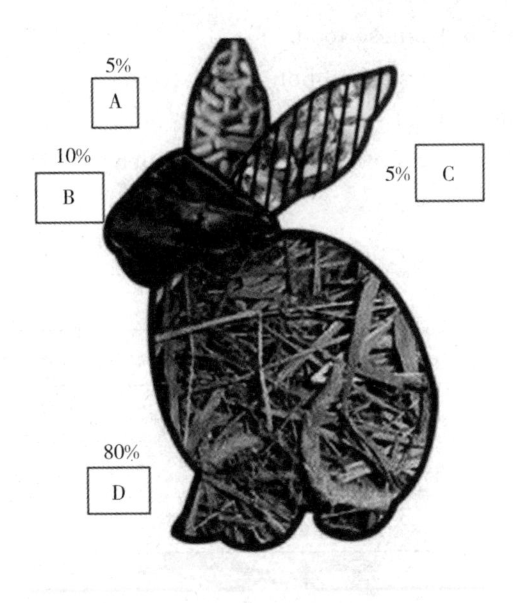

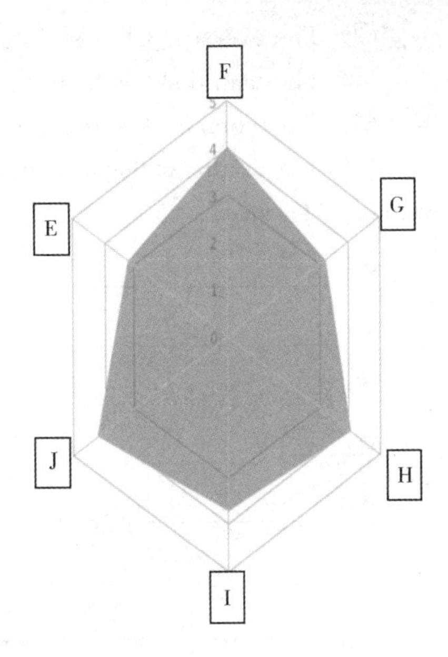

2. The second picture is a Cow-Compass. It is the risk profile of a dairy farm. Management points that may influence the quality of the milk, and the manner in which the milk is produced are assessed by six aspects, try to think what they are, and write down.

E. ________________ F. ________________ G. ________________

H. ________________ I. ________________ J. ________________

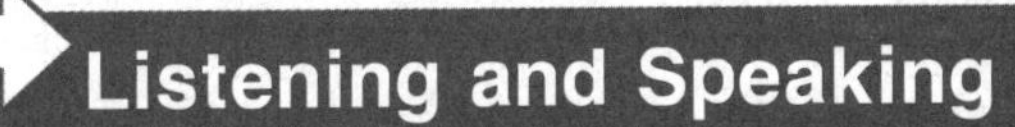

Listening and Speaking

Words and Expressions

authentic [ɔːˈθentɪk]
adj. 真的，真正的；可信的
retriever [rɪˈtriːvə(r)]
n. 寻猎物犬
formula [ˈfɔːmjələ]
n. 公式，准则；方案
absorb [əbˈsɔːb]
vt. 吸收（液体、气体等）
instruction [ɪnˈstrʌkʃn]
n. 用法说明

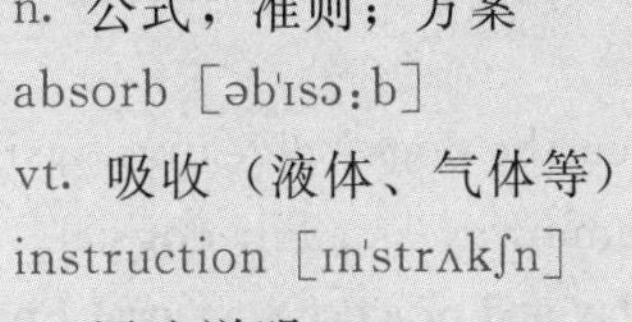

Part A Complete the following tasks according to the instructions.

1. Listen to the dialogue and judge whether the information you hear are true (T) or false (F).

(1) The woman likes the boiled beef in chili soup. ( )

(2) Sichuan food in the US is different from that in China. ( )

(3) The overseas Chinese food is as real Chinese food. ( )

(4) The man dislikes Sichuan food because it's too hot. ( )

(5) The woman likes bread very much. ( )

2. Listen to the dialogue again until you can write down the sentences (W: woman, M: man).

W: ______________________________

M: ______________________________

W: ______________________________

M: ______________________________

W: ______________________________

M: ______________________________

W: ______________________________

M: ______________________________

3. Work in pairs. Ask and answer the following questions.

(1) What kind of cooked beef was mentioned in the dialogue? Do you like spicy food? Why?

(2) How do you cook beef?

Part B Finish the following tasks.

1. Listen to the dialogue and judge true (T) or false (F).

(1) The man knows how to feed a dog. ( )

(2) Big dogs and small dogs have the same nutritional needs. ( )

(3) Digestive health is important during the puppy's growth period. ( )

(4) According to the dialogue: you can give the dogs as much food as they like. ( )

(5) When the dogs' digestive system is fully developed they can absorb nutrients in the same way as adult dogs. ( )

2. Role-play and practice your oral English. Suppose you want to raise 100 dairy cows, how would you feed them? What kinds of nutrients and feed need to be considered?

The Scientific Feeding of Dairy Cows

The scientific feeding of dairy cows mainly focuses on three factors.

First: the fertility. The health and fertility of cows are very important on a dairy farm. Healthy cows produce more milk, are more fertile, and therefore live longer. Fertile cows have a short calving interval and do not need multiple insemination to get pregnant. A sire can pass along health and fertility to his offspring. If the breeding value is above 100, the offspring will be more fertile, have better health, etc. A few examples of breeding values are: udder health, fertility, viability and calving ease.

Second: nutrition. Good nutrition means the provision of sufficient energy, proteins, minerals and vitamins. Providing a well-balanced ration does not only result in more milk, but also in better reproductive performance. In early lactation, when the milk production is at its peak, it is very hard to adjust the daily dry matter intake to the nutrient requirements of the cow, especially a high-yielding cow.

A cow's dry matter intake develops slowly during early lactation, and as a consequence an energy deficit per day is common at this time. If the diet of the cows does not contain sufficient green roughage or contains a high level of by-product feeds, deficiencies of vitamin A, phosphorus, copper, cobalt, iodine and/or selenium may arise. This may cause problems in high-yielding cows. It is important that cows continuously have access to good quality minerals of the required composition.

Feeding rations with sufficient and good quality roughage, and formulated for correct levels of protein, energy, minerals, vitamins and trace elements will normally result in a short period between calving and first heat.

Third: hygiene. Good hygiene, especially around calving is essential. Cleaning of the cow's vulva, birth-ropes and your hands before the calving process and having a clean, disinfected pen for the cow to calve will normally be sufficient.

If these things are neglected, uterine inflammation may occur. It affects the cow's subsequent fertility and it will take longer before the uterus is ready for another pregnancy. Endometritis can be diagnosed by a white mucus discharge from the vulva. It can be treated by a veterinarian, but on the other hand, the uterus may also clean itself naturally when the cow returns into heat.

Your tasks

1. Write down all words you do not know in the passage. Check the pronunciations of all, read them aloud.

____

____

____

2. Read the passage fluently and check your understanding

(1) ________ cows produce more milk, are more fertile and live longer.

(2) A few examples of breeding values are: ________, fertility, viability and ________.

(3) Good nutrition means the provision of sufficient ________, ________, minerals and ________.

(4) Good hygiene, especially around ________ is essential.

(5) Cleaning of the cow's ________, birth-ropes and your ________ before the calving process.

3. What is the principle for feeding the animals? Please share something with each other.

____

____

____

宾语和表语及其从句

1. 宾语从句

在句子中起宾语作用的从句称为宾语从句。

（1）宾语从句的语序。宾语从句的语序用陈述语序。

例句：Do you know how much the house is?

你知道这房子多少钱吗？

（2）宾语从句的引导词。

引导词	例句
由 that 引导	I know that you are a good man. 我知道你是个好人。
由 if 或 whether 引导	I don't know if he can come. 我不知道他是否能来。
由连接代词 who，what，whose 引导	Did you find out who stole the money? 你查明了是谁偷的钱吗？
由连接副词 when，why，where，how 引导	The granny doesn't know where the bus stop is. 那位老奶奶不知道公共汽车站在哪里。

2. 表语从句

表语从句是在复合句中作表语的名词性从句，放在系动词之后，一般结构是“主语＋系动词＋表语从句”。可以接表语从句的系动词有 be，look，remain，seem 等。

例句：The news is that our team got the first place.

表语从句

连接词	例句
that / whether /as if /as though（if 不引导表语从句）	The question is whether we can rely on him. 问题是我们是否能够依赖他。
连接代词：who / whom / whose / which/ what	Jane is no longer what she was four years ago. 简已不是四年前的样子了。
连接副词：when / where / why / how / because	That's because at that time we were in need of money. 那就是为什么我们在那时需要钱。 That's why I was late. 那就是为什么我晚了。

Exercises

1. Determine which types the following sentences are.

(1) Everyone knew what happened and that she was worried.

(2) The reason lies in that she works harder than the others do.

(3) Let's see whether we can find out some information about Jersey cows.

(4) I asked him where I could get that much money.

(5) The problem is that the water in our farm is not clean.

2. Analyze the underlined sentences in the following passage and then translate the paragraph into Chinese.

The animal will only realize what is going on at the moment that the veterinarian is standing before them in their white coat. How do you then use specific behavioral traits? If you know how an animal behaves in a specific situation you can take this into account. The animal responds to people as people approach it.

Coordinating human behavior with animal behavior, this is what is involved in animal handling.

____

____

____

____

____

3. Translate the following sentences into English.

(1) 那就是奶牛为什么减产。

____

(2) 问题是现在能否确定疾病的病因。

____

(3) 你知道钙盐对动物身体的重要性吗?

____

(4) 当动物受到应激时，会突然降低生产力。

____

(5) 乳制品市场状况已不同于几年前。

____

There are two passages followed by five questions, respectively. Read them and then circle the correct answer A , B, C or D for questions 1-10.

Some animals remain and stay active in the winter. They must adapt to the changing weather. Many make changes in their behavior or bodies. To keep warm, animals may grow new, thicker fur in the fall. In the case of weasels and snowshoe rabbits, the new fur is white to help them hide in the snow.

Food is hard to find in the winter. Some animals, like squirrels, mice and beavers, gather extra food in the fall and store it to eat later. Some, like rabbits and deer, spend winter looking for moss, twigs, bark and leaves to eat. Other animals eat different kinds of food as the seasons change. The red fox eats fruit and insects in the spring, summer and fall. In the winter, it can not find these things, so instead it eats small rodents.

Animals may find winter shelter in holes in trees or logs, under rocks or leaves, or underground. Some mice even build tunnels through the snow. To try to stay warm, animals like squirrels and mice may huddle close together.

Certain spiders and insects may stay active if they live in frost-free areas and can find food to eat. There are a few insects, like the winter stone fly, crane fly, and snow fleas, that are normally active in winter. Also, some fish stay active in cold water during the winter.

Some animals "hibernate" for part or all of the winter. This is a special, very deep sleep. The animal's body temperature drops, and its heartbeat and breathing slow down. It uses very little energy. In the fall, these animals get ready for winter by eating extra food and storing it as body fat. They use this fat for energy while hibernating.

Some also store food like nuts or acorns to eat later in the winter. Bears, skunks, chipmunks, and some bats hibernate.

1. The passage mainly discusses about ________ .

A. the adaptation of some animals to the changing weather

B. why many animals make changes in their behavior or their bodies

C. how animals store so much food in autumn in preparation for winter

D. how animals keep alive in the winter

2. Weasels and snowshoe rabbits are examples ________ .

A. showing food is hard to find in winter

B. belonging to the same species of animal

C. telling people that they have no difficulty in winter

D. showing how animals make changes in their behavior or bodies

3. In the winter, animals like squirrels and mice may crowd close together ________ .

A. to remain and stay active

B. to keep warm

C. because they cannot find shelter for themselves

D. because they do not need to hibernate

4. Some spiders and insects can stay alive in winter because ________ .

A. they don't need much energy

B. they can find food to eat in some frost-free areas

C. they can live inside of the trees

D. they don't have to hibernate

5. Which of the following statements are incorrect according to the passage? ________

A. The last paragraph is different in meaning from the other ones because hibernation of animals is mentioned.

B. During hibernation, the animal's body temperature drops, and its heartbeat and breathing slow down. It uses very little energy.

C. In the autumn, animals like bears get ready for winter by eating extra food and storing it as body fat. They use this fat for energy while hibernating.

D. Some animals hibernate in the winter because it is a special, very deep sleep.

Sleep is very ancient. In the electroencephalographic (脑电图的) sense we share it with all the primates and almost all the other mammals and birds, it may extend back as far as the reptiles.

There is some evidence that the two types of sleep, dreaming and dreamless, depend on the life style of the animals, and that predators are statistically much more likely to dream than prey, which are in turn much more likely to experience dreamless sleep. In dream sleep, the animal is powerfully immobilized and remarkably unresponsive to external stimuli. Dreamless sleep is much shallower, and we have all witnessed cats or dogs cocking their ears to a sound when apparently fast asleep.

The fact that deep dream sleep is rare among prey today seems clearly to be a product of natural selection, and it makes sense that today, when sleep is highly evolved, the stupid animals are less frequen-tly immobilized by deep sleep than the smart ones. But why should they sleep deeply at all? Why should a state of such deep immobilization ever have evolved? Perhaps one useful hint about the original function of sleep is to be found in the fact that dolphins and whales and aquatic mammals in general seem to sleep very little.

There is, by and large, no place to hide in the ocean. Could it be that rather than increasing an animal's vulnerability the function of sleep is to decrease it? Ray Meddis of University of London has suggested this to be the case. It is conceivable that animals that are too stupid to be quiet on their own initiative are, during periods of high risk, immobilized by the implacable arm of sleep. The point seems particularly clear for the young of predatory animals. This is an interesting notion and probably at least partly true.

6. Which of the following might be the best title for this passage? ________

A. Evolution of Sleep

B. Two Types of Sleep

C. The Original Function of Sleep

D. Animals and Sleep

7. Predators are ________ .

A. able to prey even when they are in deep dream sleep

B. more likely to experience dream sleep

C. incapable of preying when immobilized by dreamless sleep

D. good at preying on stupid animals

8. The example of dogs and cats in the second paragraph is intended to ________ .

A. explain which animals are mammals

B. show the differences between mammals

C. illustrate how shallow dreamless sleep is

D. reveal how smart they are

9. Compare with dreamless sleep, deep dream sleep is ________.

A. not the result of natural selection

B. less likely to appear to primates

C. more protective to the animals

D. at a higher stage of evolution

10. According to some scientists' research findings, dolphins seldom sleep because ________.

A. of their stupidity

B. of their vulnerability

C. there are possible dangers in the ocean

D. aquatic mammals do not need sleep

Writing Practice

You are an expert on scientific farming and a farmer asks you for advice on feeding chickens for eggs. Please make an instructional list for the chicken farmer.

Vocabulary

scientific [ˌsaɪənˈtɪfɪk]
adj. 科学的；有学问的

fertility [fəˈtɪləti]
n. 生产力，繁殖力

calving interval [ˈkɑːviŋ ˈintəvəl]
产犊间隔

multiple insemination
[ˈmʌltipl inˌsemiˈneiʃən]
多次输精

pregnant [ˈpregnənt]
adj. 怀孕的；孕育着…的

sire [ˈsaɪə(r)]
n. 雄性牲畜

udder [ˈʌdə(r)]
n. （牛、羊等的）乳房

viability [ˌvaɪəˈbɪləti]
n. 生存能力，发育能力；生活力

sufficient [səˈfɪʃnt]
adj. 足够的；充足的；充分的

protein [ˈprəʊtiːn]
n. 蛋白（质）；adj. 蛋白质的

mineral [ˈmɪnərəl]

n. 矿物；矿石；矿物质；汽水
adj. 矿物的，似矿物的
vitamin [ˈvɪtəmɪn]
n. 维生素
performance [pəˈfɔːməns]
n. 表演；演技；表现；执行
lactation [lækˈteɪʃn]
n. 哺乳期
intake [ˈɪnteɪk]
n. 吸入，摄入，摄取；纳入（数）量
high-yielding [ˈhaɪjˈɪəldɪŋ]
adj. 高产的
consequence [ˈkɒnsɪkwəns]
n. 结果，成果；[逻] 结论；重要性；推论
deficit [ˈdefɪsɪt]
n. 不足额；赤字；亏空；亏损
roughage [ˈrʌfɪdʒ]
n. 粗饲料，粗粮，粗糙的原料
deficiencies [diˈfiʃənsiz]
n. 缺乏（deficiency 的名词复数）；不足；缺点；缺陷
phosphorus [ˈfɒsfərəs]
n. [化] 磷；磷光体
copper [ˈkɒpə(r)]
n. 铜；铜币；紫铜色
cobalt [ˈkəʊbɔːlt]
n. [化] 钴（符号为 Co）；钴类颜料；深蓝色
iodine [ˈaɪədaɪn]
n. 碘
selenium [səˈliːniəm]
n. 硒
continuously [kənˈtɪnjʊəslɪ]
adv. 连续不断地，接连地；时时刻刻
composition [ˌkɒmpəˈzɪʃn]
n. 成分
formulate [ˈfɔːmjuleɪt]
vt. 构想出，确切地阐述；用公式表示
vulva [ˈvʌlvə]
n. 阴户；女阴；孔
disinfected [dɪsɪnˈfektɪd]
v. 除去（感染），给…消毒（disinfect 的过去式和过去分词）
neglected [nɪˈglektɪd]
adj. 被忽视的
v. 疏忽（neglect 的过去式和过去分词）；忽略；遗漏；疏于照顾
uterine inflammation [ˈjuːtəraɪnˌɪnfləˈmeɪʃn]
子宫炎症
subsequent [ˈsʌbsɪkwənt]
adj. 后来的；随后的；附随的
endometritis [ˌendəʊmɪˈtraɪtɪs]
n. 子宫内膜炎
diagnose [ˈdaɪəgnəʊz]
vt. 诊断；判断
vi. 做出诊断
mucus discharge [ˈmjuːkəs disˈtʃɑːdʒ]
白带呈黏性

Disease Prevention and Control

Upon completing this unit, you will be able to

◆ *identify the English words or terms related to disease prevention and control for livestock and poultry*;

◆ *master basic attributive clause and appositive clause*;

◆ *communicate in relevant activities.*

Warming up

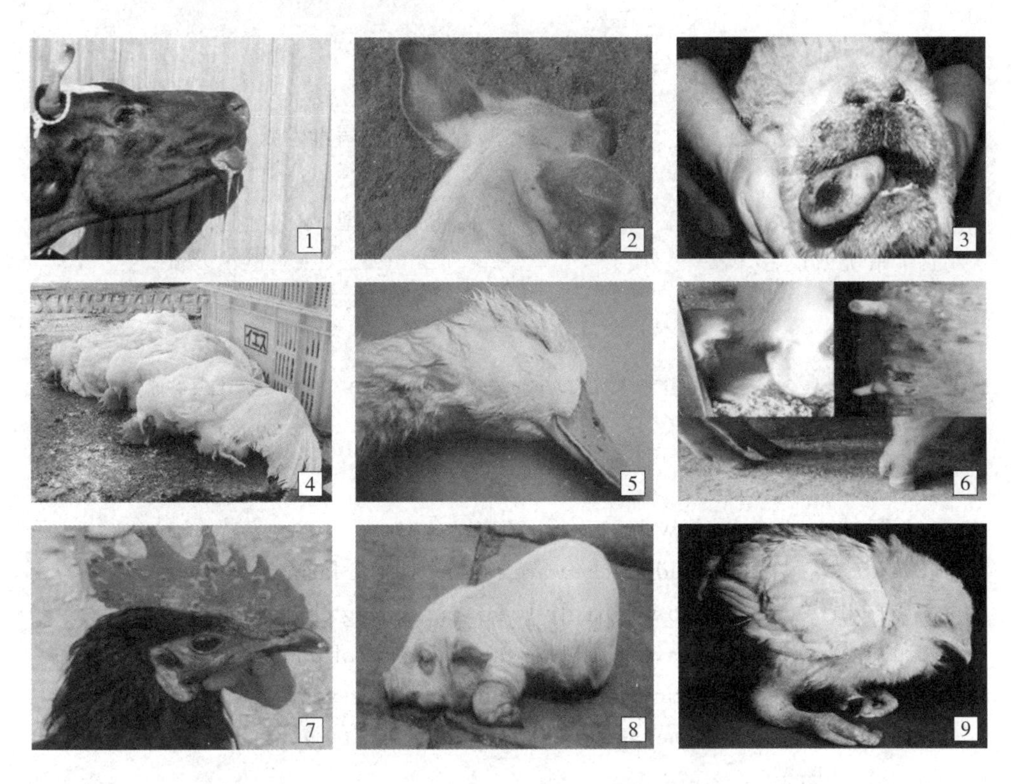

1. Do you know them? What diseases are they in English?

(1) ______ (2) ______ (3) ______

(4) ______ (5) ______ (6) ______

(7) ______ (8) ______ (9) ______

2. Match the Chinese diseases terms in column A with its translation in column B.

A	B
(1) 口蹄疫	a. Bluetongue
(2) 蓝耳病	b. Duck pestilence
(3) 鸭　瘟	c. Blue-ear pig disease
(4) 蓝舌病	d. The bird flu/Avian influenza
(5) 禽流感	e. Foot-and-mouth disease
(6) 禽　痘	f. Avian pox
(7) 猪水肿病	g. Deformity caused by a deficiency of riboflavin
(8) 核黄素缺乏症	h. Edema disease of pig

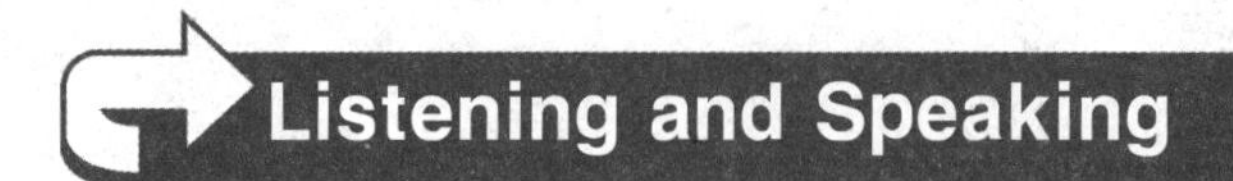

Listening and Speaking

Words and Expressions

outbreak [ˈaʊtbreɪk]
vi. 爆发
sanitation [ˌsænɪˈteɪʃn]
n. 卫生系统或设备
virus [ˈvaɪrəs]
n. 病毒；病毒性疾病
subtype [ˈsʌbtaɪp]
n. 图表类型
contaminate [kənˈtæmɪneɪt]
vt. 弄脏，污染；损害，毒害

Part A　Complete the following tasks according to the instructions.

1. Listen to the dialogue and write T for True or F for False according to the information you hear.

(1) The chicken farm was built in 2010. (　)

(2) About 150 thousand chicken are raising in the farm in all. (　)

(3) Poultry diseases do not easily break out on big chicken farm. (　)

(4) The farm owner have good measures to control poultry diseases. (　)

(5) The best way to control poultry diseases is prevention. (　)

2. Listen to the dialogue again until you can write down the sentences (W: woman, M: man).

W: ______________________________

M: ______________________________

W: ______________________________

M: ______________________________

W: ______________________________

M: ______________________________

W: ______________________________

3. Work in pairs. Ask and answer the following questions.

(1) What do you know about the common poultry diseases?

(2) What does poultry disease prevention involve?

Part B　Finish the following tasks.

1. Listen to the dialogue and answer the following questions.

(1) What is H7N9?

(2) Can the new strain of H7N9 infect humans?

(3) How do people get infected with the flu virus?

(4) Is there a vaccine to protect against this new H7N9 virus?

(5) What basic measures should people take to prevent getting infected with the viruses?

2. Do role-play and practice your oral English. Suppose you are consulting a poultry expert about H7N9.

Reading

Blue-ear Pig Disease

Porcine reproductive and respiratory syndrome virus (PRRSV) is a virus that causes a disease of pigs, called porcine reproductive and respiratory syndrome (PRRS), also known as blue-ear pig disease, which comes from the fact that infected pigs can temporarily develop discolored ears. It is one of the most economically significant diseases facing the swine industry today. PRRS is a devastating disease of pigs worldwide. It was first reported in 1987 in the United States and three years later it appeared in Western Europe and spread quickly. The disease costs the United States swine industry around $644 million annually, and recent estimates in Europe found that it costs almost 1.5b € every year.

The symptoms of PRRS include late-term fetal death, abortion, weak pigs, and severe respiratory disease in young pigs. At one-time it was called mystery pig disease. It is now known to be caused by a virus. Pigs that have been weakened by the virus are more likely to get bacterial infections. There is no cure for PRRS although a vaccine is available. Outbreaks can occur even in herds that have been vaccinated. A genetic test has now been developed that can differentiate between the harmless strain of the virus found in the vaccine and the actual disease-causing virus. Following good management practices such as biosecurity to control disease will help reduce the incidence of PRRS in a producer's herd.

The World Organization for Animal Health maintains that disease happens in most major pig-producing areas of the world. For example, an outbreak of blue-ear pig disease once killed as many as one million pigs in a year in China. The government of China says much progress has been made in efforts to control the spread of blue-ear pig disease through vaccinations and mass culls of infected pigs.

Experts say Porcine Reproductive and Respiratory Syndrome (PRRS) is a complex disease and Modified Live Vaccines (MLV) are the primary immunological tool for its control. They also verify that the disease does not seem to affect animals other than pigs and they do not know of any cases of humans who have gotten the pig disease.

Your tasks

1. Write down all words you do not know in the passage. And check out the pronunciations of all, read them loudly.

____

____

2. Read the passage fluently and check your understanding

(1) The main idea of this passage is __________.

(2) The name for the virus comes from the fact that infected pigs can temporarily develop __________.

(3) The symptoms of PRRS include __________, abortion, weak pigs, and severe respiratory disease in young pigs.

(4) The disease was first recognized in 1987 in ____________.

(5) Experts say the disease does not seem to affect animals other than ________.

3. What other pig diseases can you identify? Please give information about their symptoms.

____

____

____

Grammar

定语及其从句

1. 定语的定义及用法

(1) 定语的定义。定语是用来修饰、限定、说明名词或代词的品质与特征的句子成分。定语和中心语之间是修饰和被修饰、限制和被限制的关系。

(2) 定语的用法。

a. 用来修饰名词或代词。

b. 单个词常在被修饰的词前，短语或句子在被修饰的词之后。

c. 充当定语的词有数词、名词、形容词、副词、不定式、动名词、介词短语、从句等。

2. 定语从句的定义及用法

(1) 定语从句的定义。在复合句中，修饰某一名词或代词的从句称为定语从句。定语从句放在先行词的后面。引导定语从句的词有关系代词 that，which，who（宾格 whom，所有格 whose）和关系副词 where，when，why。关系代词或关系副词在定语从句中充当一个句子成分。

(2) 由关系词引导的定语从句。

关系词		先行词	从句成分	例句	备注
关系代词	who	人	主语	This is the man who helped me.	whom \ that \ which 在从句中作宾语时，在口语和非正式文体中常可省略，但介词提前时不能省略，也不可以用 that
	whom	人	宾语	The girl (whom) I met looks like Lily.	
	that	人、物	主语、宾语	The airplane is a machine that can fly.	
	which	物	主语、宾语	This is the book which you want.	
	whose	人、物	定语	The room whose window is red is mine.	
关系副词	when	时间	状语	I remember the day when (=on which) I got married.	when 可用 in \ on which 代替
	where	地点		I recently went to the town where (=in which) I was born.	where 可用 in \ on which 代替
	why	原因		The reason why (= for which) he didn't come was unknown.	why 可用 for which 代替

同位语及其从句

1. 同位语的定义及用法

(1) 同位语的定义。对句子中某一成分作进一步解释、说明，与前面的名词在语法上处于同等地位的句子成分称为同位语。

(2) 同位语用法。同位语常常置于被说明的词之后。可作同位语的词有名词、代词和从句等。

2. 同位语从句的定义及用法

(1) 同位语从句定义。在复合句中用作同位语的从句称为同位语从句。一般跟在某些抽象名词后，用以说明该名词的具体内容。

(2) 同位语从句中常用作先行词的抽象名词有：news，fact，truth，idea，advice，promise，belief，suggestion，hope，thought 等。常用 that 连词引导，这个 that 在从句中不充当任何成分，但不可省略。

Exercises

1. Determine whether the following sentences are attributive clause or appositive clause.

(1) That's a problem that we must pay special attention to.

(2) The fact that he was killed surprised me.

(3) Is he the man who sells dogs?

(4) There is no difficulty that can frighten us.

(5) Then it arose the question where we were to get the needed machine.

2. Find out the attributive clause from the underlined sentences in the following passage and then translate the paragraph into Chinese.

Rinderpest（牛瘟）can spread quickly through the air and in water which contains waste from animals with the virus. The disease was deadly in eighty to ninety percent of cases. It mainly sickened cattle and buffalo, but also other animals including giraffes, yaks and antelope.

Now, rinderpest expert John Anderson calls the end of the disease "the biggest achievement in veterinary history". Officials say they must still decide where to keep some of the virus and infected tissue for future research. Rinderpest is only the second disease that was ever declared to have been eliminated. The other disease is smallpox.

3. Translate the following sentences into English.

(1) 这种病只感染猪而不会感染其他动物。

(2) 猪蓝耳病最早在美国有报道。

(3) 她是一名兽医流行病学家，即研究动物传播疾病的专家。

(4) 病毒的名称来自于一个事实，就是染病猪的耳朵可能会暂时变色。

(5) 患此病的猪更容易被细菌感染。

There are two passages followed by five questions, respectively. Read them and then circle the correct answer A, B, C or D for questions 1-10.

Foot-and-mouth disease is still a major global animal health problem, but its geographic distribution has been shrinking in recent years as control and elimination programs have been established in more and more countries. Seven serotypes of foot-and-mouth disease virus have been identified by cross-protection and serologic tests, and they are designated O, A, C, SAT1, SAT2, SAT3, and Asia 1. At one time or another, these viruses occurred in most parts of the world, and often caused extensive epidemics in domestic cattle and swine.

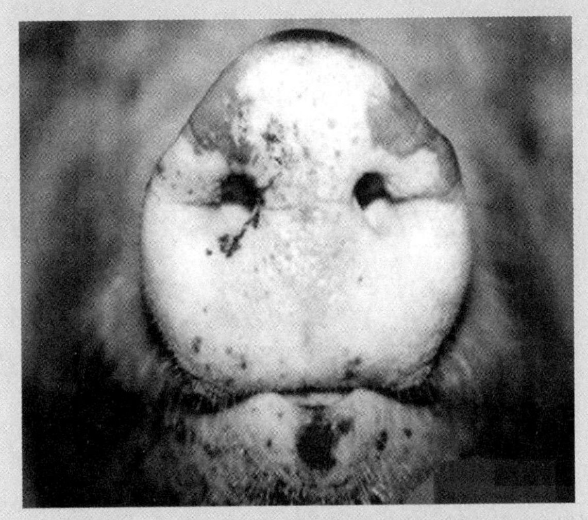

Farmers in England have been worried about foot-and-mouth disease among their cows. In recent years, foot-and-mouth disease cost the British agricultural and tourism industries billions of dollars. More than six million animals were killed. The viral sickness is one of the world's most destructive diseases of livestock. The disease affects animals such as cows, pigs, goats and sheep. It spreads easily through direct contact among animals. Foot-and-mouth disease does not usually kill animals; however, it sickens them and severely reduces their production of meat and milk, and often results in economic disaster.

1. How many serotypes of foot-and-mouth disease virus have been identified?

____

A. Five.

B. Six.

C. Seven.

D. Eight.

2. The foot-and-mouth disease virus occurred in most parts of the world, often causing extensive epidemics in domestic ________ .

A. chicken and ducks

B. cattle and swine

C. cats and dogs

D. birds

3. Foot-and-mouth disease severely reduces production of cows' ________, resulting in economic disaster.

A. appetite

B. weight

C. meat and milk

D. fertility

4. Which one of the following statements is incorrect? ________

A. Foot-and-mouth disease is not one of the world's most destructive diseases of livestock.

B. Foot-and-mouth disease cost the British agricultural and tourism industries billions of dollars.

C. Foot-and-mouth disease affects animals such as cows, pigs, goats and sheep.

D. Foot-and-mouth disease spreads easily through direct contact among animals.

5. Which title is the most accurate one for the passage? ________

A. Disease affecting pigs and cows

B. Seven serotypes of foot-and-mouth disease

C. Destructive diseases of livestock

D. Foot-and-mouth disease

It is better to prevent a poultry disease outbreak than to try to control it once it has occurred. Following the sanitation, management and vaccination suggestions will help the poultry producer prevent disease outbreaks from occurring.

The poultry flock should be checked daily for signs of disease. A sudden drop in feed and water consumption is often a sign of health problems. Watch the birds to see how they are eating and drinking. If more than one percent of the flock is infected, a disease is probably present. Death rate is another sign of disease. During the first three weeks, the normal death rate for chicks is about two percent. After three weeks of age, the death rate should not be more than one percent per month. A sudden increase in the death rate is an indication of disease.

Most poultry diseases can be accurately diagnosed only by a laboratory. Very few can be accurately diagnosed on the farm. The producer should use the services of a veterinarian to determine which disease is causing the problem. The procedure for collecting needed information and specimens is specified by the laboratory. This procedure should be carefully followed. The recommendations of the veterinarian or laboratory for control of the disease must also be followed for best results.

6. What will help the poultry producer prevent poultry disease outbreak from occurring? ________

A. Following the sanitation suggestions.

B. Following the management suggestions.

C. Following the vaccination suggestions.

D. All of the above.

7. What is often a sign of poultry health problems? ________

A. A sudden drop in weight.

B. A sudden drop in egg production.

C. A sudden drop in feed and water consumption.

D. A sudden drop in death rate.

8. How many percent of the flock is sick, a disease is probably present? ________

A. More than 1 percent.

B. About 2 percent.

C. About 3 percent.

D. About 4 percent.

9. How can most poultry diseases be accurately diagnosed? ________

A. On the poultry farm.

B. By a laboratory.

C. By a veterinarian.

D. By the poultry producer.

10. What is the best title for the passage? ________

A. Poultry Disease.

B. Poultry Disease Outbreaks.

C. Controlling of the Poultry disease.

D. Preventing& controlling Poultry Disease Outbreaks.

Writing Practice

Some purchasers who want to buy some day-old chicks will visit the poultry houses in your big chicken farm, and they need to know something about your farm and some Do's and Dont's ahead of time. Please write an announcement to them below.

Vocabulary

porcine [ˈpɔːsaɪn]
adj. 猪的，似猪的

syndrome [ˈsɪndrəʊm]
n. 综合征；综合症状；典型表现

blue-ear [bluː iə]
n. 猪蓝耳病

temporarily [tempəˈrerɪlɪ]
adv. 暂时地；临时地

economically [ˌiːkəˈnɒmɪkli]
adv. 节约地；经济地；在经济上

swine [swaɪn]
n.＜旧＞猪；＜俚＞讨厌鬼；讨厌的人

devastating [ˈdevəsteɪtɪŋ]
adj. 毁灭性的，灾难性的；可怕的

symptoms [ˈsɪmptəmz]
n. 症状（symptom 的名词复数）；征兆

fetal [ˈfiːtl]
adj. 胎儿的，胎的

abortion [əˈbɔːʃn]
n. 流产；流产的胎儿；畸形；夭折

severe [sɪˈvɪə(r)]
adj. 严峻的；剧烈的；苛刻的

bacterial [bækˈtɪərɪəl]
adj. 细菌的；细菌性

infection [ɪnˈfekʃn]
n. 传染，感染；传染病

vaccine [ˈvæksiːn]

n. 疫苗，痘苗
adj. 痘苗的，疫苗的
vaccinate ['væksɪneɪt]
vt. 给……接种疫苗
vi. 注射疫苗，接种疫苗
harmless strain of the virus
病毒无害株
disease-causing
致病性
biosecurity [ˌbaɪəʊsɪ'kjʊərətɪ]
生物安全
incidence ['ɪnsɪdəns]
n. 发生率；影响范围
herd [hɜːd]
n. 兽群；牧群；放牧人
vt. 放牧；（使）向……移动；使成群
organization [ˌɔːgənaɪ'zeɪʃn]
n. 组织；机构；团体
adj. 有组织的

n. 疫苗，痘苗
adj. 痘苗的，疫苗的
mass [mæs]
n. 大量，大多；块，堆，团
adj. 群众的；大规模的；整个的；集中的
culls [kʌlz]
n. 挑选，剔除（cull 的名词复数）
v. 挑选，剔除（cull 的第三人称单数）
infected [ɪn'fektɪd]
adj. （伤口）被感染的
v. （受）传染（infect 的过去式和过去分词）；
污染；影响
immunological [ˌɪmjʊnə'lɒdʒɪkl]
adj. 免疫学的
verify ['verɪfaɪ]
vt. 核实；证明；判定

Unit 8

Veterinary Public Health

Upon completing this unit, you will be able to

◆ *identify the English words or terms related to Veterinary Public Health (VPH);*

◆ *master adverbial clause;*

◆ *communicate in relevant activities.*

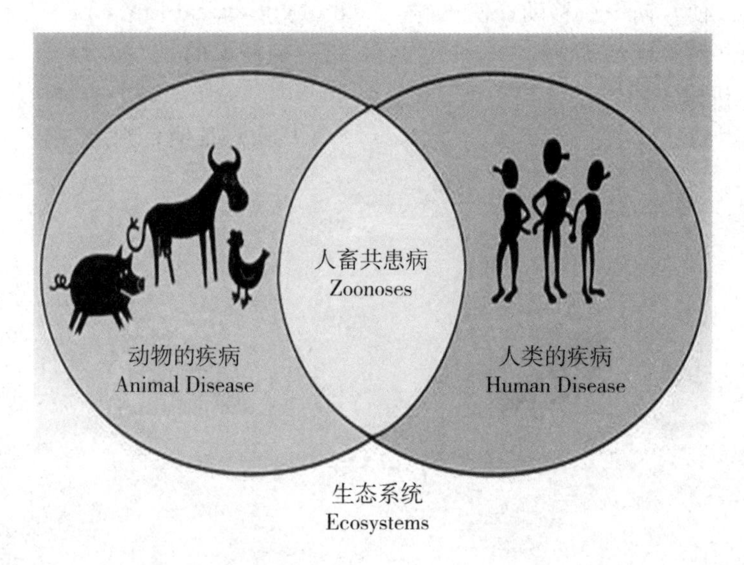

Warming up

1. Do you know them? What diseases/virus are they in English?

(1) ____________ (2) ____________ (3) ____________

(4) ____________ (5) ____________ (6) ____________

(7) ____________ (8) ____________ (9) ____________

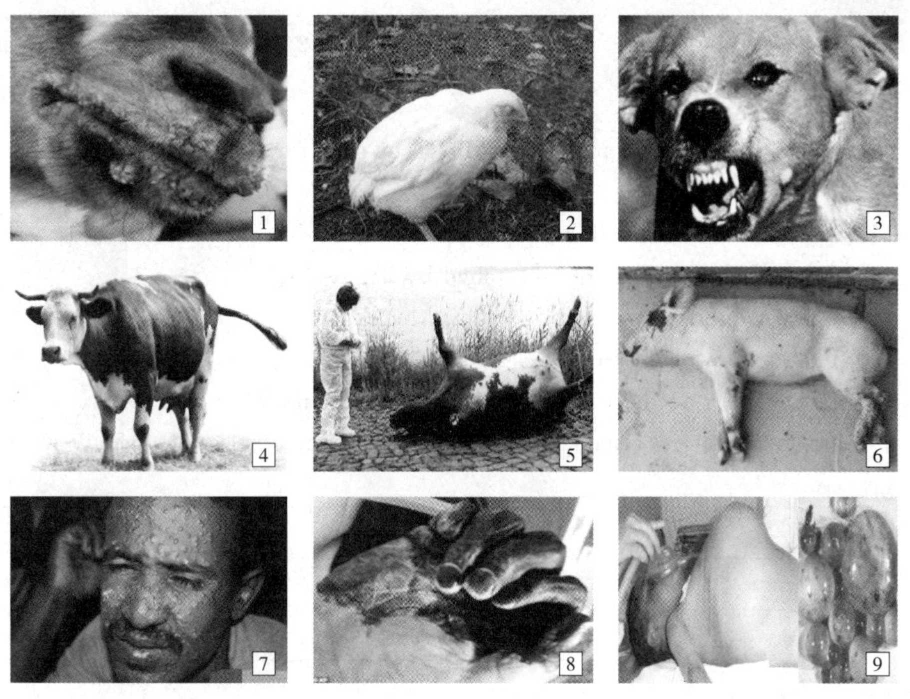

2. Match the Chinese diseases terms in column A with its English in column B.

A	B
(1) 疯牛病/牛海绵状脑病	a. Swine streptococcal diseases
(2) 布鲁氏菌病	b. Bird flu/ Avian influenza
(3) 猪链球菌病	c. Anthrax of cattle
(4) 狂犬病	d. Mad cow disease
(5) 禽流感	e. Brucellosis
(6) 牛炭疽	f. Rabies
(7) 鼠疫杆菌	g. Cutaneous Leishmaniosis
(8) 皮肤型黑热病	h. Hydatid disease of liver
(9) 肝包虫病	i. Yersinia pestis

Listening and Speaking

Words and Expressions

scratch [skrætʃ]
vt. 擦，刮；抓破
anthrax [ˈænθræks]
n. [医] 炭疽（病）
spore [spɔː(r)]
n. 孢子；胚种
undercooked [ˈʌndəkʊkt]
adj. 煮得欠熟的
preventative [prɪˈventətɪv]
adj. 预防性的

Part A Complete the following tasks according to the instructions.

1. Listen to the dialogue and write T for True or F for False according the information you hear.

(1) The woman played with her pet puppy at home when she was scratched. (　　)

(2) The woman's hand was scratched by her pet dog. (　　)

(3) The 2 wounds were shallow, only the skin was cut. (　　)

(4) The woman needs to take necessary treatments. (　　)

(5) The woman needs to get injected rabies vaccine for five times. (　　)

2. Listen to the dialogue again until you can write down the sentences(U: man, W: woman).

M: ______________________________

W: ______________________________

M: ______________________________

W: ______________________________

M: ______________________________

W: ______________________________

M: ______________________________

W: ______________________________

3. Work in pairs. Ask and answer the following questions.

(1) What do you know about rabies?

____

(2) What should you do if you are bitten by a dog some day?

____

Part B Finish the following tasks.

1. Listen to the dialogue and answer the following questions.

(1) What common zoonosis are mentioned in the dialogue?

(2) What zoonosis does the woman often hear about in her daily life?

(3) How does anthrax enter the human body?

(4) What is the best way to reduce the chance of anthrax infection in daily life?

(5) What are the good preventive measures to anthrax infection?

2. Do role-play and practice your oral English. Suppose you are consulting a professor about common zoonosis.

____

____

Reading

Zoonosis

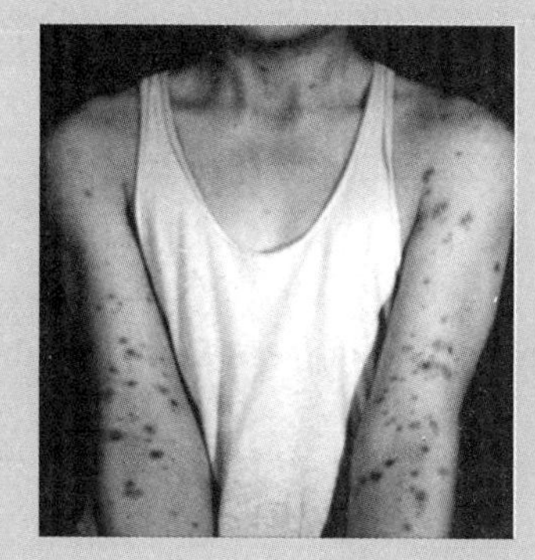

Researchers estimate that more than two billion people a year get diseases spread by animals around the world. More than two million of them die. The diseases transmitted between animals and people are called zoonosis.

A majority of human diseases are actually zoonotic. More than 60 percent of human diseases are transmitted from other vertebrate animals. Some of these diseases are pretty common. Some of the food-borne diseases and also diseases such as tuberculosis, leptospirosis are not uncommon. Others are quite rare.

Major modern diseases such as Ebola virus disease, salmonellosis and influenza are zoonosis. HIV was a zoonotic disease transmitted to humans in the early part of the 20th century, though it has now evolved to a separate human-only disease. Zoonosis can be caused by a range of disease pathogens such as viruses, bacteria, fungi and parasites; among 1,415 pathogens known to infect humans, 61% were zoonotic. Most human diseases originated in animals; however, only diseases that routinely involve animal to human transmission, like rabies, are considered direct zoonosis.

Zoonosis have different modes of transmission. In direct zoonosis the disease is directly transmitted from animals to humans through media such as air (influenza) or through bites and saliva (rabies). In contrast, transmission can also occur via an intermediate species (referred to as a vector), which carry the disease pathogen without getting infected. When humans infect animals, it is called reverse zoonosis or anthroponosis.

Here are many different infection pathways for a person. Probably the most common one is for people to get sick from food. Other transmission pathways include direct contact with animals. And some diseases can be transmitted through water or through the air.

Things could get worse in the coming years as meat production increases to feed a growing world population. High production farms often raise animals close together. Crowding can allow diseases to spread quickly. Another concern is the use of antibiotics in food animals, not only to prevent and treat diseases but to increase growth.

Your tasks

1. Write down all the words you do not know in the passage. Check the pronunciations of all and read them aloud.

____

____

2. Read the passage fluently and check your understanding.

(1) How many people get diseases spread by animals a year around the world? ________

(2) What are the diseases transmitted between animals and people called? ________.

(3) What percent of human diseases are transmitted from other vertebrate animals? ________.

(4) How many different common infection pathways for a person are mentioned in Paragraph 5 ? ________.

(5) Why are antibiotics used in food animals' production? ________.

3. What common zoonosis do you know? Please tell something about their main routes of transmission.

____

____

____

Grammar

状语及其从句

1. 状语

(1) 用来修饰动词、形容词或副词，表示动作发生的时间、地点、目的、方式等。

（2）充当状语的有副词、介词短语、不定式、分词、形容词及形容词短语、名词词组、从句等。

（3）位置比较灵活，根据需要可以放句首、句中、句末。

2. 状语从句

（1）状语从句的定义。在复合句中作状语时，起副词作用的句子即状语从句。它可以修饰谓语、非谓语动词、定语、状语或整个句子。

（2）状语从句的分类。状语根据其作用可以分为时间、地点、目的、原因、结果、条件、让步和比较等状语从句，详见下表。

类型	引导连词	例句
时间	when，whenever，while，as，before，after，since，until，as soon as 等	When I finished，I went home. Don't come in until you are called.
地点	where，wherever，everywhere	Put the book where it was.
目的	so ，so that，in order that，in case 等	He sat in the dark so that he couldn't be seen.
原因	because，as，since，for 等	We can't afford the car because it's expensive.
结果	so that，so...that，such...that	He was so weak that he couldn't walk on.
条件	If，unless，on condition that 等	If you study hard，you will pass the exam.
让步	although，even though \ if 等	Even though he is eighty，he looks strong.
比较	as \ so...as，than，no more...than	Running is not as interesting as sailing.
方式	as，just as，as if \ though 等	You must do as I told you.

Exercises

1. Try to find out the adverbial in each sentence.

(1) You will to find it where you left it.

(2) To search for gold，many people went to California.

(3) It is very kind of you.

(4) Please speak politely.

(5) Working in this way they greatly cut the cost.

2. Determine what kinds of adverbial clause are they in following sentences.

(1) Where there is a will，there is a way.

(2) Everybody likes him as he is kind.

(3) Sorry，I was out when you called me.

(4) He is ill，so he can't go to school.

(5) Whatever you say，I won't believe you.

3. Try to find out the Adverbial Clause in the following underlined sentences and then translate the paragraph into Chinese.

Swine flu is a respiratory disease caused by a combination of a virus and bacteria. Symptoms include fever, difficult breathing, coughing, going off feed, and weakness. Pigs become ill suddenly, and usually recover in about 6 days.

If swine flu becomes a problem, it may be wise to use a vaccination program to help control the outbreak. Control of the disease can generally be achieved by vaccinating the sows two times per year. After the disease is brought under control, only replacement gilts need to be vaccinated. If vaccinating the sows does not control the disease, the pigs should be vaccinated when they are seven to eight weeks of age.

____

____

____

____

____

Extra Reading

There are two passages followed by five questions, respectively. Read them and then circle the correct answer A, B, C or D for questions 1-10.

Avian influenza refers to the disease caused by infection with avian (bird) influenza (flu) Type A viruses. These viruses occur naturally among wild aquatic birds worldwide and can infect domestic poultry and other bird and animal species. Avian flu viruses do not normally infect humans. However, sporadic human infections with avian flu viruses have occurred.

Human infections with a new avian influenza A (H7N9) virus were first reported in China in March 2013. Most of these infections are believed to result from exposure to infected poultry or contaminated environments, as H7N9 viruses have also been found in poultry in China. While some mild illnesses in human H7N9 cases have been seen, most patients have had severe respiratory illness, with about one-third resulting in death. Rare, limited person-to-person spread of this virus has been identified in China, but there is no evidence of sustained person-to-person spread of H7N9.

The best way to prevent human infection with avian influenza A viruses is to avoid sources of exposure. Most human infections with avian influenza A viruses have occurred following direct or close contact with infected poultry.

People who have had contact with infected birds may be given influenza antiviral drugs preventatively. Seasonal influenza vaccination will not prevent infection with avian influenza A viruses, but can reduce the risk of co-infection with human and avian influenza A viruses. It's also possible to make a vaccine that can protect people against avian influenza viruses.

1. Which one should be the best title for the passage? ________

A. Avian influenza.

B. Avian flu viruses infect humans.

C. H7N9 viruses.

D. Spread of H7N9.

2. Avian influenza Type A viruses can infect ________ **.**

A. wild aquatic birds

B. other bird and animal species

C. domestic poultry

D. all of the above

3. When and where were first reported about human infections with a new avian influenza A (H7N9) virus? ________

A. In China in March 2016.

B. In USA in March 2013.

C. In China in March 2013.

D. In China in May 2013.

4. Most of human infections with H7N9 virus are believed to result from ________ **.**

A. exposure to infected poultry

B. contaminated environments

C. both A and B

D. none of them

5. The best way to prevent human infection with avian influenza A viruses is ________ **.**

A. to be keep away from people

B. to avoid sources of exposure

C. to be given influenza antiviral drugs preventatively

D. to be given seasonal influenza vaccination

Bovine Tuberculosis (BTB) in cattle is caused by the bacterium Mycobacterium Bovis (M. Bovis), which is a progressive wasting disease. M. Bovis is spread in a number of ways by infectious animals - in their breath, milk, discharging lesions, saliva, urine or feces. Cattle can become infectious long before they exhibit any obvious clinical signs or lesions typical of BTB, even with the most careful vet inspection. However, they can spread disease around 87 days after infection occurs. The cows that have been infected Bovine TB might lose weight and develop a cough, which spreads the bacteria through the air. Or they can appear healthy. Then, when they give birth, their calves can get infected by drinking their milk. The disease affects mainly cattle but also sheep, goats, pigs and other animals.

Control of BTB is important because of its potential zoonotic dangers and also because of the severe economic effects of slaughter and movement restrictions required to control the disease. BTB has been successfully eradicated from many developed countries including Australia, most EU Member States, Switzerland, Canada and all but a few states in the USA. EU Member states in which TB is endemic include the United Kingdom and Northern Ireland, the Republic of Ireland, Italy and Spain.

Bovine tuberculosis is a zoonotic disease and can cause tuberculosis in humans. Humans can get sick from infected cows by drinking milk that has not been heated to kill germs. Another risk is eating the meat that has not been cooked to seventy-four degrees Celsius. People who get bovine TB have to take strong antibiotics for up to nine months to cure them.

6. The cows infected Bovine TB might ________ .

A. lose weight

B. develop a cough

C. lose weight and develop a cough

D. give birth

7. When the cows infected Bovine TB give birth, their calves can get infected ________.

A. through the air

B. by drinking their milk

C. through the blood

D. by giving birth

8. Bovine TB affects mainly cattle but also ________ and other animals.

A. sheep

B. goats

C. pigs

D. all of the above

9. How long will people who get infected Bovine TB have to take strong antibiotics to cure them? ________

A. Up to nine months.

B. About one year.

C. Up to half a year.

D. Up to eight months.

10. Humans can get sick from infected Bovine TB cows by ________.

A. contacting the infected cows

B. drinking milk that has not been heated to kill germs

C. eating the meat that has not been cooked to seventy-four degrees Celsius

D. both B and C

Writing Practice

Imagine that you're a public health professional and asked to write approximately 120 words addressing the topic: what is rabies and how to reduce the risk of getting infected with rabies?

____

____

____

____

____

Vocabulary

zoonosis [ˌzuːəˈnəʊsɪs]
n. 动物病；[动] 动物传染病
transmit [trænsˈmɪt]
vt. 传输；传送，传递；发射；传染
vi. 发送信号
zoonotic [ˌzəʊəˈnəʊtɪk]
adj. 动物传染病的
vertebrate [ˈvɜːtɪbrət]
n. 脊椎动物
adj. 有脊椎的，脊椎动物的
food-borne [ˈfuːdbˈɔːn]
adj. 食物传播的
tuberculosis [tjuːˌbɜːkjuˈləʊsɪs]
n. 肺结核；[医] 结核病；痨；痨病
leptospirosis [ˌleptəʊspaɪˈrəʊsɪs]
n. 细螺旋体病
Ebola [iˈbəulə]
n. 埃博拉（病毒）
salmonellosis [ˌsælmənеˈləʊsɪs]
n. 沙门氏菌病
influenza [ˌɪnfluˈenzə]
n. 流行性感冒；[兽医]家畜流行性感冒
HIV(Human Immunodeficiency Virus)
[ˈhjuːmən ˌimjunəudiˈfiʃənsi ˈvaiərəs]
n. 人类免疫缺陷病毒，艾滋病病毒
pathogen [ˈpæθədʒən]
n. 病菌，病原体
bacteria [bækˈtɪəriə]
n. 细菌（bacterium 的名词复数）
fungi [ˈfʌŋgaɪ]
n. （fungus 的复数）真菌（如蘑菇和霉菌）
parasites [ˈpærəsaɪts]
n. 寄生物；靠他人为生的人

originate [əˈrɪdʒɪneɪt]
vt. 引起；创始；开始，发生；发明
vi. 起源于，来自；产生；起航
routinely [ruːˈtiːnlɪ]
adv. 例行公事地；常规地，惯常地
rabies [ˈreɪbiːz]
n. 狂犬病；恐水病
media [ˈmiːdiə]
n. 媒体；[解剖学] 血管中层；培养基
saliva [səˈlaɪvə]
n. 唾液，口水；津；吐沫；涎
contrast [ˈkɒntrɑːst]
n. 对比，对照；差异；对照物，对立面
intermediate [ˌɪntəˈmiːdiət]
adj. 中间的；中级的
n. 中间物；中间分子；中间人
vector [ˈvektə (r)]
n. 矢量；航向；[生] 带菌者
reverse [rɪˈvɜːs]
vt. & vi. (使)反转；(使)颠倒；交换
adj. 反面的；颠倒的；倒开的；[生] 倒卷的；
n. 倒转，反向；[机]回动；倒退；失败
anthroponosis [ænθrəˈpɒnəʊsɪs]
n. 人类疾病：人与人之间传播的疾病，
有几种可由动物传染给人（动物传染病），
有些则是在人际之间传播，特别是寄生虫病
如干皮肤利什曼病即可通过适当媒体，由一人向他人传播
pathway [ˈpɑːθweɪ]
n. 路，道；途径，路径

Unit 9

Pets

After learning this unit, you will be able to

- *gain some knowledge about pets;*
- *write an English abstracts for your articles;*
- *communicate in relevant activities.*

Warming up

1. Match the key words with the below pictures

A. goldfish ______________ B. cat ______________

C. dog ______________ D. pig ______________

E. rabbit ______________ F. lizard ______________

G. hamster ______________ H. pigeon ______________

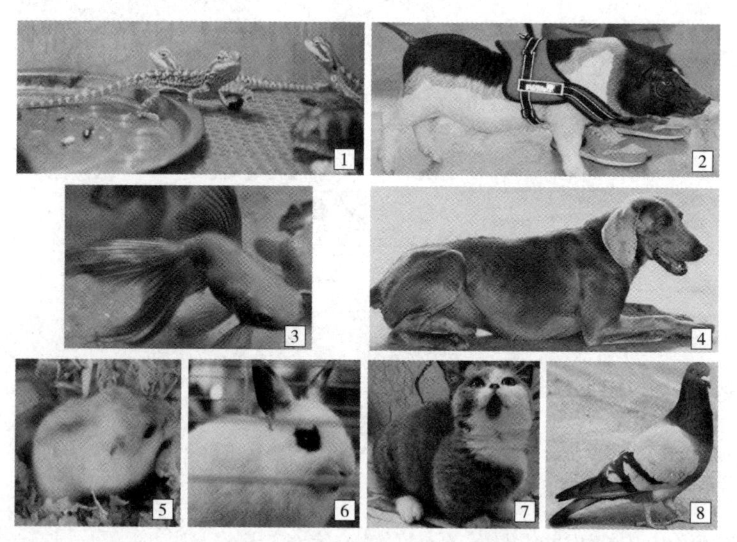

2. List any familiar pets (at least 5 species). Any pet else do you know? List them (at least 5 species).

A. ________, B. ________, C. ________, D. ________, E. ________.

more ____

Listening and Speaking

Words and Expressions

silkworm [ˈsɪlkwɜːm]
n. 蚕
low-spirited [ˈləʊˈspɪrɪtɪd]
adj. 无精神的
exotic [ɪgˈzɒtɪk]
adj. 异国的；外来的
accompany [əˈkʌmpəni]
vt. 陪伴，陪同
dung [dʌŋ]
n. 动物的粪便

Part A　Complete the following tasks according to the instructions.

1. Listen to the dialogue and judge whether the information you hear are true (T) or false (F).

(1) The boy thinks raising a goldfish is boring. (　　)

(2) The boy wants to raise a dog. (　　)

(3) The mother wants to raise a silkworm. (　　)

(4) They finally decided to raise a rabbit. (　　)

(5) The rabbit has a pair of red eyes. (　　)

2. Listen to the dialogue again till you can write down the sentences (S: son, M: mother).

S: ____________________

M: ____________________

S: ____________________

M: ____________________

S: ____________________

M: ____________________

S: ____________________

M: ____________________

S: ____________________

M: ____________________

S: ____________________

M: ____________________

S: ____________________

3. Work in pairs. Ask and answer the following questions.

(1) What pets were mentioned in the dialogue?

____

____

(2) What kind of pet would you like to have? Why?

____

____

Part B　Finish the following tasks.

1. Listen to the dialogue and fill in the boxes according to what you hear.

The benefits of keeping pets	
The disadvantages of keeping pets	

2. Do role-play and practice your oral English.

Suppose one of your friends is hesitating on whether he/she will raise a pet or not. Introduce the advantages and the disadvantages of keeping a pet to him/her to assist your friend in making decision.

What Is a Pet?

Rapid economic development has greatly reshaped our lifestyles. Raising a pet has become a new popular hobby for many people. But what is a pet? Usually, a pet or companion animal is an animal kept primarily for a person's company, protection, and/or entertainment rather than as a working animal, sport animal, livestock, or laboratory animal. Popular pets are often noted for their attractive appearances or their loyal or playful personalities. Besides animal pets, nowadays, some plants are also being planted as pets. Even some virtual pets are popular with young people.

There are various animal pets. Likely the most popular animal pets are dogs and cats. But some people also keep rodents such as gerbils, hamsters, chinchillas, fancy rats and guinea pigs as their pets. Canaries, parakeets, corvids, parrots and chickens are usually raised as pets. They are avian pets. Reptile pets, such as turtles, lizards and snakes are not strange in the flower and bird market. Aquatic pets, such as goldfish, tropical fish and arthropod pets, are common pets.

Animal pets might have the ability to stimulate their caregivers, in particular the elderly, giving people someone to take care of, someone to exercise with, and someone to help them heal from a physically or psychologically troubled past. Animal company can also help people to preserve acceptable levels of happiness despite the presence of mood symptoms like anxiety or depression. Having a pet may also help people achieve health goals, such as lowered blood pressure, or mental goals, such as decreased stress. There is evidence that having a pet can help a person lead a longer, healthier life. Having pet (s) was shown to significantly reduce triglycerides, and thus heart disease risk, in the elderly. A study by the National Institute of Health found that people who owned dogs were less likely to die as a result of a heart attack than those who didn't own one. There is some evidence that pets may have a therapeutic effect in dementia cases. Other studies have shown that for the elderly, good health may be a requirement for having a pet, and not a result.

However, not all people like to keep an animal pet. Plants are regarded as pets by more and more people who like to maintain a clean and quiet environment. Compared with pet animals, pet plants would not cause noise and excrement. Additionally, virtual pets are very popular with young people. What is a virtual pet? Actually, it is a video game. You can download the game according to your need from some specific sites and install it on your computer or cell phone. After that you get a pet. You may give a name for your pet and treat it as a real one. You need to feed it with virtual food bought from game developers when you turn on your computer. Otherwise, it would die. However, virtual pets cannot replace a real pet. It is, after all, one of the ways that game developers make money.

Your tasks

1. Write down all words you do not know in the passage. And check out the pronunciations of all, read them loudly.

____

____

____

2. Read the passage fluently and check your understanding by filling in the proper words.

(1) A pet usually is an animal kept primarily for ____________ .

(2) Rodents pets, avian pets, reptile pets, aquatic pets belong to ________ .

(3) Some research showed that ________ especially the elderly could benefit from their pets.

(4) The people who like ________ would raise a plant as their pets rather than an animal.

(5) In fact, a virtual pet is a ________ made by game developers.

3. Which pets are mentioned in this passage? Write them down.

____

____

____

____

Grammar

科技论文英语摘要写作（上篇）——英文摘要的基本结构

1. 什么是英文摘要？

将科技论文的标题、作者署名、摘要正文、关键词等四个部分翻译成英语即该论文的英文摘要。

2. 英文摘要的基本结构及撰写要点

（1）标题。英文标题应和汉语标题相对应。标题可以是短语、陈述句或疑问句，陈述句标题句尾不加标点符号，疑问句句尾要加“?”。英语标题往往先提中心词，修饰语在后，动词一般要用相应的-ing 分词、-ed 分词、不定式等。标题中的实词首字母要大写，虚词首字母一般小写。

如：Pathological Observation on Gold Fishes Infected by Parasites in Small Intestine

（2）作者署名。

①作者姓名。位于标题的下方，一般用汉语拼音表示，姓在前，名在后。对作者姓名的书写要求，不同的期刊要求不一样，多数期刊要求将姓氏的拼音全部字母大写，名的第一个字母大写。

如：GAO Deng-hui，LIU Fang，LI Chao-bo。

②作者工作单位。一般要用括号括上，用斜体字体表示。通常小单位在前，大单位在后，工作单位后面是单位所在城市、邮政编码及地区。

如：(*College of Animal Science*，*Guizhou University*，*Guiyang* 550025，*China*)。

有多个作者的，每个作者的工作单位都要写上，顺序与汉语摘要的一致。字母大小写规则同标题部分。

（3）摘要正文。

①摘要正文前加“Abstract：”。

②时态。科技文写作常用一般现在时、一般过去时和现在完成时三种时态。

③语态。写英文摘要时，往往采用第三人称的被动语态。

（4）关键词。先用“Keywords：”表示，然后将汉语摘要中的关键词翻译成英文单词，英文单词的首字母要大写，每个关键词之间要用“；”分隔，最后一个关键词后不加标点符号。

如：Keywords：Pumpkin；Effective Component；Glucatonic；Diabetic Rat Model

Exercises

1. Determine which part of an English abstract the following expressions are.

(1) Pathological Observation on a Case of Canine Fibrosarcoma

(2) ZHANG Xiao-hua①, WANG Ke-mei②, TANG Ren-long③

(3) (*College of Veterinary Medicine, Northwest Agriculture and Forestry University, Yangling 712100, China*)

(4) Abstract: The purpose of this study was to establish a canine model of knee osteoarthritis...

(5) keywords: cat; X-ray; osteoarthritis

2. Analyze the structure of the English abstract and translate it into Chinese.

Observation on the Structure of Frog's Kidney

GAO Deng-Hui① LI Chao-Bo②

(①*College of Animal Science, Guizhou University, Guiyang 550025, China*; ②*Guangxi Vocational College of Agriculture, Nanning 530001, China*)

Abstract: The microstructure of frog's kidney was studied by using histological and histochemical methods. The results indicated that beside nephrons and collecting tubes, lymphoid tissues could be seen in the frog's kidney. Additionally, a structure was found at the ventral side of the kidney, which is similar to the Corpuscle of Stannius of teleostei (真骨鱼类斯坦尼斯小体) and in which a lot of mast cells could be seen. It suggested the kidney of frog was a multifunctional organ.

Keywords: Frog; Kidney; Structure; Mast cell

3. Translate the following sentences or phrases into English.

(1) 人与宠物关系的社会学研究。

(2)(贵州农业大学动物医学院 贵阳 550001)

(3) 本文介绍了犬、猫常见皮肤病的诊断及治疗。

(4) 通过组织化学染色观察了兔子肾中肥大细胞的分布。

(5) 结果表明，该调查中犬急性胰腺炎的发病率与性别、年龄、品种有关系。

Extra Reading

There are two passages followed by five questions, respectively. Read them and then circle the correct answer A, B, C or D for questions 1-10.

One out of 15 Chinese people owns a pet, with an average pet costing 500 yuan ($74) per month. There is great potential in the domestic pet market because pet owners in China pay increasing attention to their new family members. That has created a new segment for businesses. You will find various professional services for pets such as pet products, pet behavior training, pet first aid, pet-sitting, pet care, and so on.

Pet hospitals and pet care shops are the two most common forms of pet business. There are hundreds of pet hospitals and pet care shops in some metropolises like Beijing, Shanghai, Hangzhou, Guangzhou, etc. Dozens of that, at least, could be found in some second-tier cities like Guiyang. You can even see them in many county towns. Only pet hospitals and pet care shops are not enough. Some other businesses of the pet care industry need to be developed and improved.

Supermarkets, in which, pet owners could get all of the things related to pets, will be very necessary. There are 5,000 categories of pet supplies available in some pet markets. But it is a few. Pet owners have to buy these things everywhere.

Sometimes, the offspring of pets is strongly expected by their owners. So Matchmaking Service Center is highly recommended. Pet owners could find a husband or a wife for their pet from the center. In addition, the service of wedding dress and wedding photography may be provided.

As we know, bringing a pet is forbidden by almost all restaurants. It is a problem that pet owners don't know where they should leave their pets when they go to a restaurant. Pet owners neither want to leave their pets home nor keep them out of the restaurant. The best way to solve the problem is offering dining service for pets in the same restaurant but different rooms. We may call it as pet restaurant. Then pet owners and their pets can enjoy their meal simultaneously. Pet owners no longer worry that their pets are treated coldly.

Just like humans, pets will die one day. Running a funeral parlor for pet maybe is a best choice. Pet funeral parlors are rare in China and are mostly found in major cities. All corpses of pets could be disposed by the funeral parlor after their death. It will avoid environmental pollution caused by littering corpses. Most of all, taking a burial for pet will relieve the owner's grief.

1. What is this passage mainly about? ________

A. New segment of pet businesses. B. The benefits of raising a pet.

C. The disadvantage of raising a pet. D. How to raise a pet.

2. Which pet business is not mentioned in this passage? ________

A. Pet hospital B. Funeral parlor for pets

C. Supermarket of pet supplies D. Pet clothing

3. The businesses of pet industry needed to be developed and improved except ________.

A. Pet Grooming Shop B. Matchmaking Service Center

C. Pet restaurant D. Funeral parlor for pets

4. Which of following statement is not true according to the passage? ________

A. Chinese regard their pets as their family members.

B. Pet clinic is rare in some metropolises like Beijing and Shanghai.

C. Pet owners have to buy supplies everywhere for their pets.

D. Pets can have their meal with their owners when the owners go to a restaurant but in different rooms.

5. Why the author thought running a funeral parlor for pets was a best choice? ________

A. Pets will die in the future.

B. Funeral parlor for pets is rare in China.

C. It avoids environmental pollution.

D. Taking a burial for pet in the funeral parlor will relieve the owner's grief.

The most popular pets can vary from one country to the next, but on a worldwide basis, house cats are generally thought to be more common than any other pets. Dogs are the second most popular pets, and these are generally followed by fish, rodents, birds and various reptiles. The method used to measure pets can cause this to vary because some studies look at households while others look at the actual number of animals. So, while a person may own only a single dog, some fish owners may have a dozen fish in one tank. An example of how this can **skew** statistics occurs in America, which has more households that own dogs, but has cat owners that collectively have so many individual animals that they still manage to outnumber the dogs on a country-wide basis.

Cats and dogs are both popular pets for similar reasons. They can offer personal affection for people, and they bond with their owners to varying extents. Dogs are naturally social animals that crave a pack environment, and this causes them to form an especially tight bond, while cats also form a bond that is slightly different, but still generally powerful. In addition, both animals have been useful to people from a practical perspective, with cats being used as rodent killers and dogs having many important uses, from hunting to security.

Fish are some of the most popular pets partly because people love to watch them. They're colorful and some people get a great sense of joy from seeing them swim inside their artificial environments. They're also generally low-maintenance pets that don't take up that much space.

People keep rodents as pets for slightly different reasons. Sometimes they're mostly kept in their cages, and people may purchase them for some of the same reasons they buy fish. Other people form very close bonds with their pet rodents and take them out of their cages often to pet them. Some types, like rabbits, are commonly allowed to roam a person's house without a cage much of the time, and people may treat them exactly like a cat or a dog.

Birds sing, and some types can even learn to speak. This, along with their colorfulness and easy maintenance, make them some of the most popular pets. Many bird species also have very long life spans, and this often appeals to people.

Exotic reptiles are some of the most popular pets for people with allergies because they have no fur. They also appeal to people who want to own an animal that's a little bit closer to being wild. Most of these animals aren't actually domesticated at all, and some of them can even be a little dangerous if they aren't handled exactly right. Maintenance for reptiles can sometimes be difficult because they may require very carefully controlled environments to avoid infections or problems with body temperature.

6. The most appropriate topic for this passage is ________.

A. The most popular pets　　B. All kinds of pets

C. The advantage of pets　　D. The role of pets

7. What does the word "skew" mean most likely in the first paragraph? ________.

A. wrong　　B. exact　　C. inaccurate　　D. bias

8. What caused house cats are the most popular pets? ________

A. Cats are more beautiful than other pets.

B. Some studies look at households while others look at the actual number of animals.

C. Cats can offer more personal affection for people than other pets.

D. There are more and more households that own dogs.

9. Cats and dogs are both popular pets because several reasons except ________ .

A. they can offer personal affection for people

B. they fond with their owners to varying extents

C. they are useful to people from a practical perspective

D. they can kill rodent

10. Which of following statement is not true according to the last paragraph?

____

A. Exotic reptiles are suited to be raised by people with allergies.

B. Exotic reptile appeal to people who want to own an animal that's not wild but mild.

C. Some of exotic reptile might be a little dangerous.

D. Exotic reptiles are difficult to maintain.

Writing Practice

Suppose the article below was written by you. Please make an English abstract for it.

宠物营养研究现状及发展前景

（作者是你自己）

（作者单位是你的学校）

摘要：宠物也称伴侣动物，常被视为家庭成员。随着人们生活水平的提高，对宠物犬和猫科学饲喂的要求也日益增强，宠物饲养逐步向商业化方向发展。宠物食品是经济发展到一定程度的产物，具有一些鲜明的特点。虽然现在养宠物的人越来越多，但对于宠物营养全面认识的却少之又少。本文对近年来国内外宠物饲料的研究状况进行了综述，从而为宠物营养的研究提供参考依据。

关键词：宠物；宠物营养；宠物食品；宠物饲料

____

____

____

____

____

____

____

Vocabulary

pet [pet]
n. 宠物；宠儿；受宠的人；生气
adj. 宠爱的，溺爱的；表示亲昵的
vt. 宠爱；爱抚，亲抚
reshape [ˌriː'ʃeɪp]
vt. 重塑；给…以新形态；采取新形式；打开新局面
companion [kəm'pæniən]
n. 同伴；同甘共苦的伙伴
vt. 同行；陪伴
entertainment [ˌentə'teɪnmənt]
n. 娱乐，消遣；款待；娱乐节目
laboratory [lə'bɒrətri]
n. 实验室；实验课；研究室；药厂
appearance [ə'pɪərəns]
n. 外貌，外观；出现，露面
loyal ['lɔɪəl]
adj. 忠诚的，忠心的；忠贞的
playful ['pleɪfl]
adj. 闹着玩的；爱玩的；戏谑
personality [ˌpɜːsə'næləti]
n. 生而为人；人格，人品，个性；人物
virtual ['vɜːtʃuəl]
adj. 实质上的，事实上的；虚拟的
rodent ['rəʊdnt]
n. 啮齿目动物
adj. 咬的，嚼的；啮齿目的；侵蚀性的
gerbil ['dʒɜːbl]
n. 沙鼠
hamster ['hæmstə (r)]
n. 仓鼠
chinchillas [tʃin'tʃiləz]
n. 南美洲栗鼠（chinchilla 的名词复数）
fancy rat ['fænsi 'ræt]
n. 花式鼠
guinea pig ['gɪnɪ pɪg]
n. 豚鼠，天竺鼠；试验品，实验对象
canaries [kə'neərɪs]
n. 金丝雀（canary 的名词复数）
parakeet ['pærəkiːt]
n. 长尾小鹦鹉
corvid [kɒvid]
n. 鸦科的鸟类，包括乌鸦、松鸦、喜鹊
avian ['eɪviən]
adj. 鸟的，鸟类的
reptile ['reptaɪl]
n. 爬行动物；卑鄙的人
adj. 爬虫类的；卑鄙的
turtle ['tɜːtl]
n. 龟；[动] 海龟
vi. 捕海龟，捕鳖
lizard ['lɪzəd]
n. 蜥蜴
tropical ['trɒpɪkl]
adj. 热带的；炎热的；热情的
arthropod ['ɑːθrəpɒd]
n. & adj. 节肢动物（的）
stimulate ['stɪmjuleɪt]
vt. 刺激；激励，鼓舞；使兴奋
vi. 起兴奋作用；起刺激作用
caregiver ['keəgɪvə (r)]
n. 照料者，护理者
psychologically [ˌsaɪkə'lɒdʒɪklɪ]
adv. 心理上地，心理学地

preserve [prɪ'zɜ：v]
vt. 保持，保存；腌制食物；防腐处理
vi. 保鲜；保持原味
depression [dɪ'preʃn]
n. 萎靡不振，沮丧；衰弱
significantly [sɪg'nɪfɪkəntli]
adv. 意味深长地；值得注目地
triglycerides [traɪg'lɪsəraɪdz]
n. 三酰甘油
therapeutic [ˌθerə'pjuːtɪk]
adj. 治疗的，疗法的；有益于健康的
dementia [dɪ'menʃə]
n. 痴呆
excrement ['ekskrɪmənt]
n. 排泄物（粪便）；屎

Unit 10

Animal Products

After studying this unit, you will be able to

- ◆ *understand the knowledge about livestock poultry products*;
- ◆ *write an outline of a graduation paper*;
- ◆ *write scientific and technological articles in English*;
- ◆ *communicate in relevant activities*.

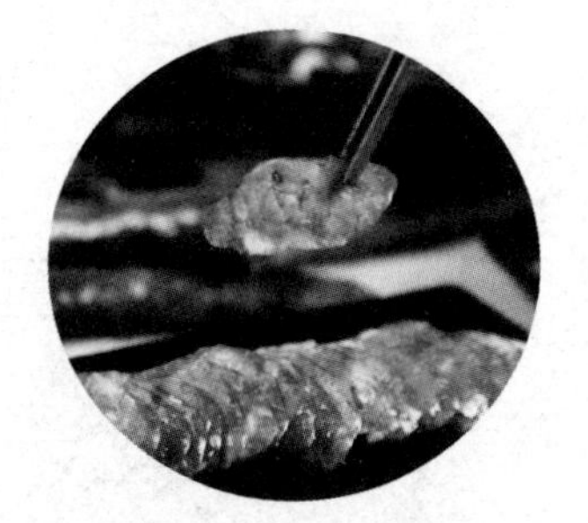

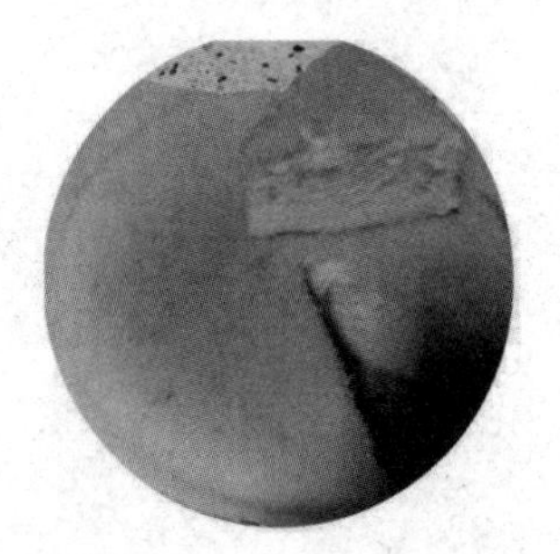

Warming up

1. How many of the livestock and poultry products do you already know? There are various of food made of them in our daily life. Organize the above pictures according to the following categories.

Meat products: ______________________________

Dairy products: ______________________________

Egg products: ______________________________

2. The blood, bone, internal organs, fur and hoof are all parts of livestock and poultry, which are called as animal by-products and can either be destroyed or used to make compost, biogas in western countries. But in China, animal by-products can be made into many useful things. How many of these useful products do you know? List them in the table below.

Blood	
Internal organs	
Fur	
Hoof	
Bone	

Words and Expressions

steak [steɪk]
n. 牛排；肉排

leftover [ˈleftəʊvə (r)]
adj. 剩余的；未用完的

refrigerator [rɪˈfrɪdʒəreɪtə (r)]
n. 冰箱；冷藏库

buffalo [ˈbʌfələʊ]
n. 水牛；野牛

camel [ˈkæml]
n. 骆驼

Part A　Complete the following tasks according to the instructions.

1. Listen to the dialogue and write T for True or F for False according to the information you hear.

(1) They are talking about their supper. (　　)

(2) The wife bought some chicken. (　　)

(3) The meat had normal smell. (　　)

(4) The husband forgot to put the meat in the refrigerator. (　　)

(5) They have some fresh eggs and pork. (　　)

2. Listen to the dialogue again till you can write it down (H: husband, W: wife).

H: ______________________________

W: ______________________________

H: ______________________________

W: ______________________________

H: ______________________________

W: ______________________________

H: ______________________________

W: ______________________________

H: ______________________________

W: ______________________________

H: ______________________________

3. Works in pairs. Tell your partner your favorite food. Do you often cook for yourself?

Part B

1. Listen to the passage and then fill in the blanks with missing words.

________ is often seen as the most complete natural food. It provides many of the ________ for the heath of the human body and is an excellent source of ________, ________ and minerals, too. ________ are not the only source of milk, because milk can be obtained from many different sources. The milk from ________ and ________ makes a great contribution to the total milk production all over the world, especially in Europe. On the contrary, the ________ is a common source of milk in many parts of ________ and milk from ________ and ________ is also drunk by some people.

2. Do role-play and practice your oral English.

Suppose a new restaurant opens nearby you. You receive news that their dishes will be on sale, so you plan to invite your good friend, Anna, to have dinner there.

____

____

Reading

Meat Products

As one of the most valuable livestock products, meat includes other bioactive components and a small amount of carbohydrates, besides protein, amino acids, minerals, fats, fatty acids and vitamins. In nutrition, the importance of meat is reflected in containing high-quality protein, amino acids and its highly bioavailable minerals and vitamins.

The consumption of meat has been relatively stable in the developed world, however, the annual consumption of meat per capita has doubled in developing countries since 1980. The demand for livestock products is gradually increasing with the growth of the populations, the people's income and the change of food preferences.

It is predicted that meat production in the world will be doubled by 2050, which will mostly happen in developing countries. The expansion of meat marketing in developing countries provides significant opportunities for livestock farmers and meat manufacturers. Nevertheless, with the increasing of animal production, it is also a serious problem for meat processing, marketing, quality and hygienic safety.

By regulating the safe, efficient and sustainable production of meat products, The Food and Agriculture Organization (FAO) aims to provide opportunities for livestock development and poverty alleviation in its member states.

The plan of the FAO aims to enhance the skills and capacities of smallholder by improving and upgrading small-scale meat production and processing techniques. With the combination both at headquarters and in the field, and the cooperation at the local, national and international levels, FAO offers assistance to market meat products and completes the value chain of meat products.

Your tasks

1. Write down all the words you do not know in the passage. Verify the pronunciations and read them aloud.

______

______

______

2. Understanding the text: answer the following questions with the information contained in the reading.

(1) According to the text, what is meat composed of?

______

(2) Why do people think meat is important from the nutritional point of view?

______

(3) What's the situation regarding meat consumption in developed and developing countries?

______

(4) Is there a big challenge for meat products? What is it?

______

(5) What is the plan of FAO?

____

3. What is a meat product according to your understanding? Please tell us something about it.

____

____

Grammar

科技论文英语摘要写作（下篇）——英文摘要的常用表达句型

科技论文摘要是论文的重要组成部分，摘要的内容包含有与论文同等量的主要信息，它包括研究内容、研究目的、研究方法、研究结果（结论）等方面的基本内容，以供读者确定有无必要阅读全文。在撰写英语摘要时，常用以下句型来表达这些基本内容。

1. 表示文章的论点或研究内容

This paper is mainly devoted to...

This report is a study of...

This paper introduces...

2. 表示研究目的的句型

This study was conducted/ intended to...

This research was aimed at...

The purpose/ main objective of this paper/ study is...

This study attempt to...

3. 表示研究方法的句型

This paper is based on...

This article provides an exploratory methodology to...

用 by+-ing 或 through+-ing 形式

4. 表示研究结果或主要发现的句型

It was found that...

The finding showed that...

The survey revealed that...

5. 表示研究结果（结论）的句型

It is concluded that...

The study suggests that...

The results suggest/ show...

The results indicated...

Exercises

1. Determine if the sentences below are the objective, the content, the method, the result or the conclusion of a research.

(1) The amino acid composition of lamb was analyzed by systematic analysis and compared with several other species in this study.

(2) The objective of this study was to discuss beef performances utilization of crossbred Simmental by crossbreeding crossbred Simmental with Black Angus.

(3) The results showed that 17 kinds of amino acids were detected in both yak milk and Holstein milk.

(4) Those results revealed the protein expression pattern difference between MFGM protein of dairy cow and goat milk.

(5) It is concluded that the breed and feeding age can affect the nutritional composition and textural properties of chicken.

2. Analyze the abstract below and decide which is the objective, the method, the result and the conclusion of the research. Then translate it into Chinese.

Abstract: This research was aimed at the effect of duck fat on the duck flavor. Triglycerides and phospholipid in duck meat was extracted selectively by using petroleum ether and chloroform with methanol respectively. Sensory analysis and instrument analysis showed that triglycerides had little effect on the duck characteristic and meaty aroma, but phospholipid was inverse. Duck fat extracted by chloroform with methanol was washed by water to remove the water soluble compounds in fat. No S-and N-containing compounds were identified in washed duck fat, and the characteristic aroma compounds were significantly lower ($p<0.05$) than the control duck fat. The water soluble compounds in duck fat contributed the duck meaty aroma but not the characteristic aroma of duck.

____

____

____

____

3. Translate the following sentences into English.

(1) 本试验对四川白鹅蛋品质进行了研究。

____

(2) 本试验旨在研究中草药制剂对鹌鹑蛋黄胆固醇含量的影响。

____

(3) 通过正交试验研究了不同条件下制备的鸡肉丸子的感官特性。

---

(4) 研究结果显示，性别对猪肉的品质特性有明显影响。

---

(5) 结论：当 pH 为 6.8 时，羊奶的热稳定性最大。

Extra Reading

There are two passages followed by five questions respectively. Read these passages and choose the best answer for each question.

Eggs are one of the important sources of nutrients for human beings. Egg white contains rich protein and yolk has abundant high-quality lecithin. With the increasing market share of eggs in our country, more and more egg products are coming on the market. Egg products refer to the semi-finished products, by-products or highly-finished products using the eggs as raw material and going through the primary and further processing. The common egg products on the market are egg white powder, yolk powder, dried egg and salad dressing.

The manufacturers who need egg products are cake shops, the catering industry and food producers. According to how the industry uses egg products, there are six usages of egg: the function of gelation in the industry of cake and cooked food; bubbling effect in the industries of making cookies, cakes and semi-finished food; the function of emulsification which uses egg product as emulgator owing to the high viscosity of yolk; the function of crystallization which is used in making cookies; being natural pigment, which means eggs, as natural pigment, are widely used and highly

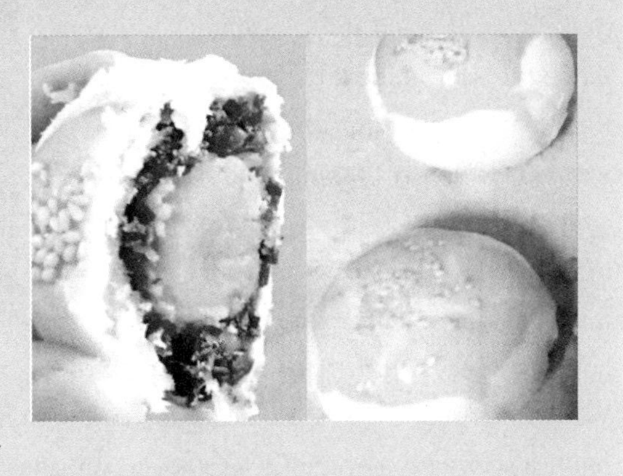

praised in making cookies, cooked wheaten food and sauce; and the last one is being an unique condiment, because of the unique egg fragrance, yolk and whole egg play a supporting role of seasoning.

1. What's the best title of this article? ________

1. Egg powder　　B. Yolk　　C. Egg product　　D. Eggs' function

2. Which of the following in (is) NOT true? ________

A. Egg white contains rich protein.

B. Egg yolk has abundant high-quality lecithin.

C. There is a great market share of eggs in our country.

D. There are more egg products than eggs in market.

3. The manufacturers such as ________ need egg product.

A. food producers　　B. fabrics industry

C. catering industry　　D. cake shops

4. What's not the function of egg product? ________

A. Gelation　　B. Emulsification

C. Crystallization　　D. Germination

5. Based on the passage, it can be assumed that ________ .

A. egg plays a very important role in the food industry

B. yolk can be used in every food production field

C. egg products are very popular for young people

D. egg white is generally used in salad dressing

Which is healthier, whole milk or skimmed milk?

The number of the people who prefer low-fat milk or skimmed milk is not low because in their mind milk without fat or containing less fat is healthier and at the sa-me time better for weight loss, especially in hot weather. However, skimmed milk has excluded much unhealthy saturated fat, thus lowering the cholesterol level and calories in milk but the amount of protein is not changed; protein is the main nutrition that we hope to obtain from the milk, explained by the national secondary public dietitian Want Zhe. Once fat is excluded from milk, it will not only become tasteless because butyric acid with a special fragrance in milk is excluded, but fat-soluble vitamin such as vitamin A, B, E and vitamin K will be significantly reduced and the content of liquid-soluble vitamin and minerals will also be reduced. Even more regrettably, CLA (conjugated linoleic acid) in milk is excluded

in the process of skimming fat, which is beneficial to anticancer and because of redu-cing the content of vitamin D in the process of skimming fat, the calcium absorption rate are also affected although the content of calcium in milk is not reduced too much in the process. In conclusion, fat-skimmed milk and low-fat milk lose much important nutrients in the process of skimming fat.

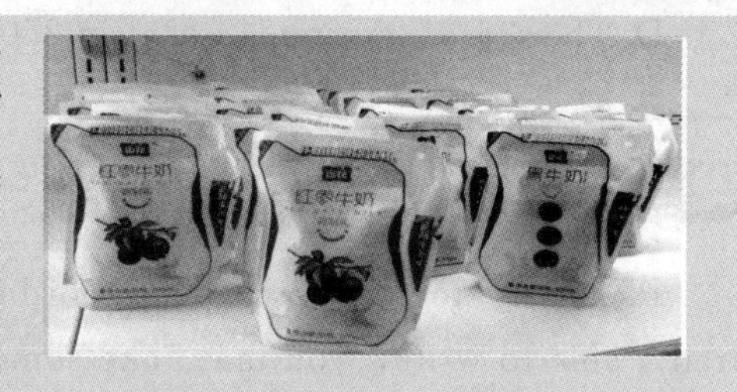

In fact, compared with the high-fat food like animal innards and fat meat, 4% of fat in the whole milk is not a big amount. Based on the consumption level of 250 milliliter per person every day, someone who drinks whole milk will consume 7.5 grams more fat than someone who drinks skimmed milk. The 7.5 grams of fat amounts to the fat content of one spoonful of cooking oil, or a single piece of fatty meat. Therefore, unless it is aimed at individuals who need to strictly control their diet, such as people with diabetes, cardiovascular patients or the those suffering from obesity and abnormal metabolism, all other healthy people, especially teenagers should prefer whole milk.

6. According to the Want Zhe's explain, whole milk________ .

A. is less healthy than skimmed milk

B. holds less cholesterol

C. tastes better

D. excludes saturated fat

7. Which is beneficial to anticancer? ________

A. CLA　　B. Vitamin A, B, E　　C. Vitamin D　　D. Butyric acid

8. Who is not proposed (supposed) to drink skimmed milk? ________

A. Diabetes

B. Teenagers

C. Cardiovascular patients

D. Abnormal metabolism people

9. If you drink 250 milliliters whole milk it is equal to________ .

A. a piece of fat meat

B. two spoon of cooking oil

C. a cup of coffee

D. a spoon of soup

10. What's TRUE in the following sentences? ________

A. There is more fat in whole milk than innards.

B. While milk belongs to high-fat food.

C. Skimmed milk loses many important nutrients in the process of skimming fat.

D. There are 7. 5 grams of fat in 250 milliliter skimmed milk.

Writing Practice

Please try to write a paper according to your major in English. In case you aren't able to write, you may take some notes from your reading so that you can master more useful sentences and the grammer structures included in the article

____

____

____

____

Vocabulary

bioactive [ˌbaɪˈæktɪv]
adj. 对活质起作用的，生物活性的
component [kəmˈpəʊnənt]
n. 成分；零件；[数] 要素；组分
adj. 组成的；合成的；构成的
carbohydrates [kɑ：bəˈhaɪdreɪts]
n. 糖类（ carbohydrate 的名词复数）；
淀粉质或糖类食物
amino [əˈmiːnəʊ]
adj. 氨基的
acid [ˈæsɪdz]
n. 酸（acid 的名词复数）；酸味物质
fatty acid [ˈfæti：ˈæsid]
n. 脂肪酸
reflect [rɪˈflekt]
vt. 反射，反照；表达；显示；折转
vi. 映出；反射；深思熟虑；慎重表达
bioavailable [baɪəˈveɪləbl]
adj. 生物及生物药效应的
consumption [kənˈsʌmpʃn]
n. 消费；肺病；耗尽；消耗性疾病
annual [ˈænjuəl]
adj. 每年的；一年的；[植物] 一年生的
capita [ˈkæpitə]
n. 头，首（caput 的复数）
preference [ˈprefrəns]
n. 偏爱；优先权；偏爱的事物
nevertheless [ˌnevəðəˈles]
adv. 然而；尽管如此；不过；仍然
conj. 然而；尽管如此
poverty [ˈpɒvəti]
n. 贫穷；不足；贫瘠，不毛；低劣
alleviation [əˌli：vɪˈeɪʃn]
n. 减轻；缓解；缓和；镇痛物
enhance [ɪnˈhɑ：ns]
vt. 提高，增加；加强
capacities [kəˈpæsitiz]
n. 容量（ capacity 的名词复数 ）；才能；性能；生产能力
smallholder [ˈsmɔ：lhəʊldə（r）]
n. 小农，小佃农
upgrading [ˈʌpgreɪdɪŋ]

n. 浓缩
v. 提升（ upgrade 的现在分词 ）
technique [tek'niːk]
n. 技巧；技能；技术；技艺
combination [ˌkɒmbɪ'neɪʃn]
n. 结合；联合体
headquarters ['hed'kwɔːtə]
n. 总部（pl.）
assistance [ə'sɪstəns]
n. 帮助，援助
predict [prɪ'dɪkt]
vt. 预言，预测；预示，预告
vi. 预言，预示
expansion [ɪk'spænʃn]
n. 扩大；扩张；扩张物；膨胀物

参考答案

Unit 1

Warming up

1. Grassland，sheds，slope and mountain.
2. Grassland：A，I
   Shed：B，C，D，E，H
   Slope and mountain：F，G

Listening and Speaking

Part A

1. (1) F　(2) F　(3) T　(4) F　(5) T
2. 详见本单元参考答案后听力材料。
3. 略

Part B

1. A
2. A. tall　B. black and white　C. short hair　D. milk
3. B

Reading

1. 略
2. 例：

(1) 参考第一段：It is a major part of agriculture and one of the important material-producing departments. It's a kind of way to get great economic value by breeding livestock in captivity，and then obtain animal by-products or working animals.

(2) 参考第二段：Australia-New Zealand Mode，Europe Intensive Mode，and Traditional Mode.

(3) 参考第二段：Europe Intensive Mode，mainly relies on the capital input；and Traditional Mode，its main characteristic is labor input.

(4) 参考第四段：Chinese animal husbandry has been experiencing the transition from traditional husbandry to modernized husbandry.

(5) 开放性答案：Sure，it will become successful.

3. variation，by-products，Industrialization，adopted，referred to

译文：

畜 牧 业 经 济

畜牧业作为农业的重要组成部分和重要的物质生产部门之一，它是以动植物的生产能力为基础，通过圈养家畜获得动物副产品或役用动物，创造经济价值。畜牧业中，牲畜分为两类：一类是肉用动物类，如鸡、马、牛等，另一类是商品类，如鹿、熊、貂等。现代畜牧业不仅为人类提供肉、蛋、奶和血，而且还为轻工业提供毛、皮和骨等原材料。同时，畜牧业还通过粪便的利用和动物役用功能为种植业提供无法估量的帮助。

全世界范围内，不同的国家根据土地、资本和劳动力的不同特点，因地制宜采取不同的畜牧业经济发展模式。例如，以草地投入为主的澳大利亚、新西兰采用草地畜牧业发展模式，简称澳新模式；欧洲集约化模式以资本投入为主，而传统模式是劳动力投入为主。中国过去主要采用传统的畜牧业发展模式。然而，由于人口众多，草原面积缩小，耕地减少，农场数量减少，导致近几十年来我国畜牧业正经历着由传统畜牧业向现代化畜牧业的转变。又由于草地资源质量差、地域差异大、经济发展不平衡，中国畜牧业逐渐适应了多种形式。这些形式包括：个体农户、专业农户、大农场和大规模饲养区。从总体上看，农户家庭经营模式仍然是畜牧业生产的主要形式。随着我国工业化和城市化进程的不断推进，个体农户模式日益向大农场模式发展。由于中国畜牧业的特点，中国畜牧业正不断向欧洲集约化、规模化模式发展，并将在不久的将来逐步实现现代化。

Grammar

Exercises

1. (1) 主语　(2) 谓语　(3) 状语　(4) 定语　(5) 宾语

2. 画线部分成分依次为：状语，主语，状语，定语，表语。

译文：

随着中国经济的发展，各行各业都在摸索一条正确的道路来应对中国面临的严峻挑战。就中国畜牧业的发展而言，存在着两种明显的发展模式：一种是草地畜牧业和天然草场上的山地畜牧业，另一种是舍饲畜牧业。根据我国的自然条件，草地畜牧业和天然草地上的山地畜牧业一直是我国畜牧业的主要部分。近年来，随着科学技术的发展，舍饲畜牧业也呈现出蓬勃的发展趋势。

3.

(1) One of the by-products of potato is potato chips.

(2) The report never referred to the developing situation of Chinese animal husbandry economy.

(3) Our farm adopted new breeding skill to face social development.

(4) What are the specialties of Chinese animal husbandry?

(5) The animal husbandry is thriving and prospering day by day.

Extra Reading

1. A 2. B 3. D 4. B 5. C 6. C 7. C 8. B 9. A 10. D

Writing Practice

Personal Information Form

<table>
<tr><td>Name</td><td>Meimei, Han</td><td>Gender</td><td>Male \ Female</td><td>Date of Birth</td><td>09/28/1997</td><td rowspan="4">Photo</td></tr>
<tr><td>Student ID</td><td>NO：××</td><td>Nationality</td><td>CHINA</td><td>Politics status</td><td></td></tr>
<tr><td>Department</td><td>Economic management</td><td>State of health</td><td>Healthy/ feel well/ good enough</td><td>Major</td><td>E-commerce</td></tr>
<tr><td>Address</td><td colspan="5">Eg：No. 32，××Road，××District，××City，×× Province.</td></tr>
<tr><td>Graduate school</td><td colspan="2">×× High School</td><td>Hobby</td><td colspan="3">Eg：Reading，dancing...</td></tr>
<tr><td colspan="2">Phone NO.</td><td colspan="2">××</td><td colspan="2">E-mail</td><td>××</td></tr>
<tr><td colspan="2">Family member and family contact</td><td colspan="5">Father：×× Tel：××
Mother：×× Tel：××</td></tr>
</table>

扩展：

Politics status：communist party member/ league member or member of CYL/ the masses or public people or citizen/ personage or patriots without party affiliation

听力材料

Part A：

Emi：Hi，I am Emi. I live in Apartment 23.

Guo：Hello，Emi. You can call me Guo. I'm in 18. Nice to meet you.

Emi：Nice to meet you too.

Guo：Are you a student，Emi?

Emi：Yeah，I'm majoring in animal husbandry and veterinary science.

Guo：Wow，that sounds interesting.

Emi：How about you，Guo?

Guo：I am also a student in the agricultural university，I study ecological agriculture.

Emi：Aha，we can study together!

Part B：

Teacher：Hello? Guo! Please wake up!

Guo：Hi，Miss Li，I'm so sorry，I worked late last night.

Teacher：Really? Can you tell me what Holstein cattle look like?

Guo：Holstein? It is a breed of cattle which is tall，yellow，and it has long brown hair.

Teacher：Oh，OK! Does it wear glasses?

Guo：Ah，yes.

Teacher：Seriously? Cattle wear glasses?

Guo：Sorry，Miss Li.

Teacher：OK，tell me your correct answer.

Guo：Holstein is a kind of dairy cattle，black and white cow，and milk usually come from these cattle.

Teacher：Not too bad!

Unit 2

Warming up

1. A. chicken B. horse C. goose D. cattle E. pig F. duck G. sheep H. rabbit I. deer

2. Livestock：B、D、E、G、H、I

Poultry：A、C、F

Listening and Speaking

Part A

1. (1) T (2) F (3) F (4) F (5) T

2. 详见本单元参考答案后听力材料。

3.

(1) Horse，sheep and cattle.

I like all of 3 species. However，I like horses most. There are a lot of reasons：

①Horses are creatures of beauty.

②Riding is all around great exercise. Burning calories and toning muscles while enjoying nature is great. If we ride regularly we can probably save a bundle on gym fees.

③Horses relax us，they not only bring us pleasure，but can provide emotional support. They seem to sense depression and pain in a person. They greet us with whinnies, and nuzzle us when we are feeling sad, and always have two ears to listen to our problems, and a strong shoulder to cry on.

Therefore, incorporating horses into your life is a good way to create a healthy living habit.

(2) At present, I am still a student. I hope to become the owner of a farm. Maybe, you can tell me what I should do now?

Part B

1.

			Peking Ducks			
1000	600	1600		0	3050	Alipay

2. 略

Reading

1. 略

2. (1) Heritage breeds (2) heritage (3) supplemental feed (4) modern improved breeds (5) intended purpose or use

3. 例：

Peking duck is an excellent domestic duck heritage breed. This breed has the bird characteristics that include pure white feathers, fecundity, fast growth, succulent fatty meat and so on. The raw materials of the most famous dish "Beijing Roast Duck" is **Peking Duck.**

Chinese Holstein cow, named "Black and White Dairy Cattle" before 1992, is the only dairy breed in China. This breed was formed after having bred for a long time by hybridizing a variety of Holstein cows and local cattle.

Duroc pig is an older breed of American domestic pig that forms the basis for many mixed-breed commercial hogs. Duroc pigs are red, large-framed, medium length, and muscular, with partially drooping ears, and tend to be one of the least aggressive of all the swine breeds.

译文：

畜 禽 品 种

畜禽品种是指通过选种繁育，使其变得彼此相似，并将这些相似的性状均衡稳定地传给他们的后代的家畜群。

世界上，所有种类的家畜和家禽几乎都有不同的品种。从育种的角度来看，它们包括适合小规模农庄的遗产品种和适合大规模作业的现代改良品种。

什么是遗产品种？

关于“遗产”畜禽品种，没有国际上的官方的定义或认证，然而，它们通常被当作是历史性的品种，是由我们的祖先养育的存在了几十年甚至数百年的品种。它们都通过精心选种和培养，随时间推移慢慢地发展其性状，这些性状使得它们非常适应当地环境。它们是在与现代农业截然不同的耕作方式和文化条件下发展起来的。

遗产品种的显著特点是味道好，但个体小。例如：香猪、金华猪、北京鸭。与现代牛种相比，传统的牛种通常生产高品质肉和奶，它们以草为食加上少量的补充饲料就能茁壮成长。同样，传统的鸡品种与现代快速生长的鸡品种相比，传统鸡的产蛋率高、肉质坚韧，但通常需要更长的时间才能完全成熟。

现代改良品种

现代改良品种主要是指杂交品种和外来品种。用在大规模的农业中的许多品种是为了集约化生产而专门选育的。其中包括生长速度、饲料转化率、连续产奶或产蛋量，或其他有针对性的生产特征。如：为了瘦肉多，我们选择约克夏和/或杜洛克猪；为了产蛋量，选择白来航鸡；为了产奶量，人们选择荷兰荷斯坦牛。

总之，不同的动物品种选择取决于预期的目的或用途。

Grammar

1.

(1) They raised 10,000 geese in the farm. 简单句（主语＋谓语＋宾语结构）。
主语 谓语 宾语 状语

(2) China is a developing country; America is a developed country.
并列句（两个分句都是主语＋系动词＋表语结构）

(3) He offered me his seat. 简单句（主语＋谓语＋间宾＋直宾结构）

(4) In order to catch up with the others, I must work harder. 简单句（主语＋谓语结构）

(5) You must work hard if you are afraid of failing. 复合句，主句是（主语＋谓语结构）

2.

The pig was one of the earliest domesticated animals in China.
主语 系动词 表语

此句属于简单句（主语＋系动词＋表语结构）

The improved pig breeds of ancient Rome, Britain and USA（主语）, which have had a tremendous influence on pig breeding in nearly every part of the world（非限定性定语从句）, **had been influenced**（谓语）by breeding stocks imported from southern China（状语）.

因为含有从句，此句属于复合句。主句属于主语＋谓语结构

译文：

在中国，猪是被最早驯化的动物之一。养猪在中国有着悠久的历史，中国的繁育技术技艺十分精湛，从而培育出了数量众多的优质猪种。中国为世界提供了相当

数量的最珍贵的猪种。古罗马、英国和美国的改良猪种对世界各地的养猪业产生了巨大的影响，这些改良品种本身也受到从中国南方进口的猪种的影响。中国猪的品种数量庞大，为我国和其他地区的家畜进一步改良提供了丰富的遗传资源。

3.

(1) White Leghorn has high ability of laying eggs.

(2) They are going to send us 100 sheep.

(3) There are 90 Holstein cows on our farm.

(4) Duroc is a pig with high lean meat rate.

(5) People think the Hampshire pig is a very good meat variety.

Extra Reading

1. D 2. A 3. D 4. C 5. B 6. A 7. B 8. A 9. C 10. D

Writing Practice

Dear Mr. Henry

I am Tom from the small town of the New Investment and Development Zone, not very far from your place. I have been building a family farm that will be finished before the end of next month. We have a need for about 200 goats, 50 cows, and 5 oxen. I wonder what breeds you can offer. Would you kindly tell me the approximate price for each breed?

I look forward to hearing from you soon. Thank you.

Regards,

Tom Jones

听力材料

Part A

W: What is your job on the farm?

M: My job is to feed the animals, such as the horses and the sheep.

W: What breeds of sheep live on the farm?

M: Well, there is the Texel sheep which is a hardy breed of sheep.

W: Can I see them, please?

M: Certainly! There is a picture of the Texel sheep on the wall.

W: Wow! It looks like an ox.

M: Really? In fact, it is much gentler than others.

Part B

W：Hi，can I help you?

M：Yes. I would like to buy some young poultry.

W：What kinds of poultry would you like?

M：I would like chickens，ducks and geese.

W：We have several different breeds.

M：What are they?

W：Well，we have many breeds，for example：the white Leghorn，which lays a good number of eggs，the New Hampshire，which gains weight and feathers quickly and produces large brown eggs，the Peking duck，which is one of the most popular breeds of ducks and has a fast growing rate and good hatching ability. I am sorry we have no geese at the moment.

M：OK. 1000 White Leghorns，600 New Hampshires，500 Peking ducks. How much will that cost?

W：Just a moment，let me check. $3,050 in total.

M：OK. Can I use Alipay or WeChat to pay for them?

W：Yes，both are okay.

Unit 3

Warming up

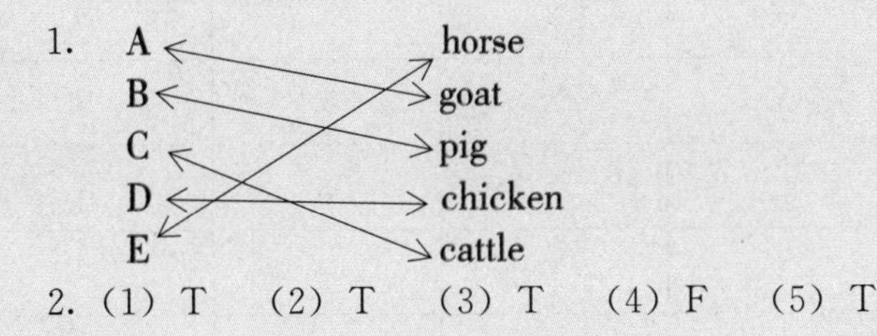

2. (1) T (2) T (3) T (4) F (5) T

Listening and Speaking

Part A

1. (1) F (2) F (3) F (4) T (5) T

2. (1) serious (2) thinking about (3) anatomized (4) consist (5) surprised (6) anatomy

3.

A：What does this picture illustrate?

B：Well. This is a ruminant stomach. The stomach has four parts，they are rumen，reticulum，omasum and abomasum. Because of this four-part stomach， digestion in ruminants differs from that in non-ruminants.

Ruminants eat rapidly. They do not chew much of their feed before they swallow it through the esophagus. The solid part of the feed goes into the rumen. The liquid part goes into the reticulum, to the omasum, and then to the abomasum and on into small intestine.

When the rumen is full, the animal lies down. The feed is then forced back into the mouth and rumination occurs.

When the animal eats large amounts of fresh grass or legumes, a large amount of carbon dioxide and methane gas can be formed by bacterial action in the rumen. These gases must be disposed of through the digestive system. Otherwise, bloating will take place if the gases form faster than the animal can eliminate them. Most of the gases are discharged when burping.

Part B

1.

Types of animals	Ruminants	Non-ruminants	Herbivores	Carnivores	Omnivores
Swine	N	Y	N	N	Y
Dogs	N	Y	N	Y	N
Sheep	Y	N	Y	N	N
Goat	Y	N	Y	N	N
Cats	N	Y	N	Y	N
Cattle	Y	N	Y	N	N
Horses	N	Y	Y	N	N
Poultry	N	Y	N	N	Y

2. 略

Reading

1. 略

2.

(1) F (The key point is the word "like", it means "the same as...", it is not used as a verb.)

(2) F (The systems of poultry are composed of some organs.)

(3) T　(4) T　(5) T

3.

The anatomy of the livestock and poultry refers to the science that deals with the form and structure of the animals. Physiology handles the study of functions of the body or any of its parts. A thorough knowledge of the structure of an animal imparts a lot of information about the various functions that it is capable of performing, as well, assist us in recognizing the normal, in order to determine the abnormal; help us understand how to diagnose disease or determine if an animal has sustained an injury; help understand the physical capabilities or limitations of particular species; understand what happens in the nutrition and growth processes; and assist us to get better performance from the animals.

This knowledge is essential if you want to work with the animals in any capacity.

译文：

畜禽解剖学概述

家畜和家禽，跟所有高等生物一样，都是由细胞组成的。它们从单个细胞（受精卵或卵子）开始，发展成多细胞的有机体。随着细胞的分裂和生长，它们分化为具有多种功能的组织。

动物体由不同种类的组织组成。这些组织聚集起来形成器官。再由器官形成系统，如：骨骼系统、肌肉系统、循环系统、呼吸系统、神经系统、内分泌系统、泌尿系统、消化系统和生殖系统。每个系统都在动物体内执行一些重要功能。

骨骼系统包括骨骼、软骨、牙齿和关节。它给予身体形状、保护、支撑和力度。动物体内有三种肌肉：骨骼肌、平滑肌和心肌。肌肉系统帮助身体运动并履行其他重要功能，如维持心脏跳动。心脏是循环系统的一部分。心脏、动脉、毛细血管和静脉是循环系统的主要组成部分。心脏起到水泵的作用，使血液在全身流动。血液将氧气和营养物质输送到所需的细胞中，并将细胞中的废物去除。呼吸系统把空气吸入体内，氧气在体内被提取，氧化，为动物体提供能量。携带废气的空气，如二氧化碳，就会被排出体外。主要的呼吸器官有：鼻腔、气管、支气管和肺。神经系统提供了一种方法，使细胞能够根据需要将信号从身体的一个部位传送到另一个部位。神经系统的两个主要部分是中枢神经系统和周围神经系统。内分泌系统分泌生长和身体发育所需的激素。泌尿系统将一些废物排出体外。泌尿系统的主要成分是肾、输尿管、膀胱和尿道。消化系统（或道）由身体中与咀嚼和消化食物有关的部分组成。即：口腔、食道、胃、小肠、大肠、直肠、肛门。由于消化器官结构的不同，不同动物的消化能力差异很大。生殖系统是指促成有机体产生后代的性器官。

生物体的生存依赖于所有器官系统的综合活动，这些系统通常由内分泌和神经系统协调。

Grammar

1.

(1) taste→tasty 作为并列关系的表语，用词要统一，small 是形容词，所以，要把 taste 变成形容词形式。

(2) wonderfully→wonderful 通常用形容词修饰名词。

(3) informations→information 不可数名词不能加“s”或“es”。

(4) see→sees 主语第三人称单数，一般现在时，谓语动词要加“s”或“es”。

(5) cost→take。It takes somebody to do something 是固定句型。

2.

The reproductive system of the female chicken is in two parts: the ovary
形容词（定语） 介词词组（表语）
and oviduct. Unlike most female animals, which have two functioning ovaries,
介词词组（状语） 现在分词（定语）
the chicken usually has only one. The right ovary stops developing when the
名词（主语） 动名词（宾语）
female chick hatches, but the left one continues to mature.
动词（谓语） 不定式（宾语）

The ovary is a cluster of sacs attached to the hens back about midway
过去分词词组（定语）
between the neck and the tail. The oviduct is a tube like organ lying along the
现在分词词组（定语）
backbone between the ovary and the tail.

译文：

母鸡生殖系统分为卵巢和输卵管两部分。与其他雌性动物不同，大多数雌性动物有两个功能正常的卵巢，鸡通常只有一个卵巢。雌性雏鸡孵出时，右侧卵巢停止发育，而左侧卵巢继续生长发育直到成熟。

卵巢是附着在母鸡背上的一簇囊，在脖子和尾巴之间。输卵管像一个管状器官，附着在卵巢和尾巴之间的脊椎上。

3.

(1) Our knowledge of animal anatomy is crucial for us to scientifically raise animals.

(2) I learned many important details about dissecting in today's lab class.

(3) The digestive system (or alimentary tract) consists of the body parts used in chewing and digesting food.

(4) The reproductive system of birds is different from that of mammals.

(5) “Rumen bubbling” occurs frequently in herbivores such as cattle and sheep.

Extra Reading

1. B 2. D 3. C 4. A 5. C 6. A 7. B 8. D 9. D 10. D

Writing Practice

Heart: Arterial blood (oxygen-rich blood) flows from the heart to each part of the body to provide oxygen and nutrients. The venous blood (oxygen-poor blood) returns from the body to the heart. The blood then travels through the lungs to exchange carbon dioxide for new oxygen. The heart is a pump, which moves the blood. The arteries and veins are the pipes through which the blood flows. The lungs provide a place to exchange carbon dioxide for oxygen.

Stomach: The main function of the stomach is to chemically and mechanically break down food. It accomplishes this by secreting stomach acid and enzymes to digest food and churning the food by the periodic contraction of the stomach muscles.

Lung: Lung is an important part of the respiratory system. The lungs play a critical role in the body, extracting oxygen from inhaled air for distribution via the bloodstream to every cell in the body. Conversely, during exhalation the lungs expel waste--carbon dioxide produced when cells use oxygen.

Liver: The liver's main functions are:

A. bile production and excretion;

B. excretion of bilirubin, cholesterol, hormones, and drugs;

C. metabolism of fats, proteins, and carbohydrates;

D. enzyme activation;

E. storage of glycogen, vitamins, and minerals;

F. synthesis of plasma proteins, such as albumin, and clotting factors;

G. blood detoxification and purification.

听力材料

Part A

H: Hi Lisa. Why do you look so serious?

L: Oh, I am just thinking about the class I attended this morning.

H: What class?

L: It is an anatomy class. During the class, we anatomized one pig and one goat.

H: Wow, two types of animals. Could you tell me something about them?

L: Sure! As far as I could see, they both consist of hair, skin, bone, muscle and body fluid. However, there was one thing that surprised me. The goat had quite a big stomach with 4 compartments, despite having a small body.

H: Oh, I know. Goats are ruminants. Tomorrow I will also have an anatomy class during which we will dissect a cow.

L: I bet you are looking forward to this class.

H: Yes, really.

Part B

W: Hello

M: Hi. This is Tom. Can I speak to Mary?

W: Hi Tom. This is Mary speaking. How are you today?

M: I am fine. I am thinking about having a picnic tomorrow. Would you like to come with us? So far it will be Jerry and me.

W: Oh, I'd love to, but I have to drive to another town and buy chicken feed.

M: There is a shop selling feed across the road. Why not buy there?

W: Oh, only pig fodder is being sold there. Different animals require different kinds of feed because of different structures of gut organs. Cattle, sheep and goats are ruminants, and they need only grass. Horses are non-ruminants, but also herbivores because their large intestines are exceptionally large and complex, similar to the rumina of ruminants. Poultry and pigs are both omnivores, but they need different feed with different nutrients.

M: What kinds of feed do dogs and cats need?

W: Well, they are both carnivores, so they primarily need animal fodder.

M: Thank you for telling me. See you later.

W: It is my pleasure. See you.

Unit 4

Warming up

1. A. cell　B. cell nuclear　C. chromosome　D. chromatin　E. protein　F. gene　G. basic group　H. sperm　I. egg

2. It is fertilization. 顺序为：(5) → (4) → (2) → (1) → (3)

Listening and Speaking

Part A

1. (1) T　(2) T　(3) T　(4) F　(5) F

2. 详见本单元参考答案后听力材料。

3.

(1) Check the bull semen quality.

(2) The experiment is to test the bull semen quality.

(3) 略。

Part B

1.

(1) chicken slaughter measurements

(2) live-weight, carcass weight, semi-eviscerated weight, eviscerated weight, breast meat rate, leg meat rate, abdominal fat percentage

(3) 8

(4) 32

2. 例:

This week, I have learned the Artificial Insemination (AI) experiment, mainly on cows. AI is the deliberate introduction of sperm into a female's cervix or uterus, the purpose of which is to achieve a pregnancy. AI is done through in vivo fertilization by means other than sexual intercourse. AI is a common practice in animal breeding, including dairy cattle (see Frozen bovine semen) and pigs. It is also a fertility treatment for humans. I find this experiment is very important to our future career, and it needs a variety of practice. So I will practice as much as I can to improve my skill.

Reading

1. 略

2. (1) genetically (2) variation (3) genotype (4) phenotype (5) interplay

3. 例:

ABO blood type

Parents	Genotype	Children
O×O	OO×OO	O
O×A	OO×AO	O, A
	OO×AA	A
O×B	OO×BO	O, B
	OO×BB	B
O×AB	OO×AB	A, B

（续）

Parents	Genotype	Children
A×B	AO×BO	O，A，B，AB
	AO×BB	B，AB
	AA×BO	A，AB
	AA×BB	AB
A×A	AO×AO	O，A
	AO×AA	A
	AA×AA	A
B×B	BO×BO	O，B
	BO×BB	B
	BB×BB	B

译文：

基因型和表型

了解基因型和表型

19世纪末至20世纪初，丹麦的一位科学家 Wilhelm Johannsen 通过一系列实验，观察到了遗传上相同的豆子有不同的变异。因此，他断定：这个变化必然是由环境因素造成的，并在1911年首次提出了“基因型”和“表型”这两个术语。

基因型是每个有机体的遗传组成。基因型的作用就好像是一个指令集，为个体的生长和发育提供一系列的指令。当人们谈论某一特定性状（如眼睛颜色）的遗传时，通常用“基因型”这个词。

表型是某个有机体的可观察的物理或生化特征，由遗传组成和环境因素共同决定，例如身高、体重和肤色。

基因型如何影响表型

术语“基因型”通常用于表示特定的等位基因。等位基因是同一基因的两种形式，在染色体上占据同一位置。在任何给定的位点，有两个等位基因（对每一个染色体上的两个等位基因），一个来自父亲，一个来自母亲。两个等位基因可能是相同的，也可能是不同的。一个基因的不同等位基因通常具有相同的功能（例如，它们编码的蛋白质影响眼睛的颜色），但可能会产生不同的表型（例如，蓝色的眼睛或棕色的眼睛），这取决于你拥有哪两个等位基因。

例如，PTC（苦味化合物）尝味能力是由一个单一的基因所控制。该基因至少有7个等位基因，但其中只有2个是常见等位基因。

一个大写的“T”代表了显性的等位基因，具有PTC尝味能力，“显性”意味着任何人有1个或2个显性等位基因T就能够品尝出PTC。味盲等位基因是隐性基因，由一个小写的“t”表示。“隐性”意味着一个人需要2份的隐性等位基因才表现为 PTC 味盲。

每一对等位基因代表一个个体的特定基因型，在这种情况下，有 3 种可能的基因型：TT（味尝）、Tt（味尝）和 tt（味盲）。如果等位基因是相同的（TT 或 tt），基因型是纯合的。如果等位基因不同（Tt），基因型是杂合的。

总之，基因型或基因构成对生长发育起着至关重要的作用。然而，在整个生命中，环境因素始终影响表型，这就是遗传和环境之间持续的相互作用，使得所有个体都是独一无二的。

Grammar

1.

(1) Mr. Hu is not only a businessman but also a teacher. 并列句

(2) China is a developing country; America is a developed country. 并列句

(3) Changsha is the place where I was born. 定语从句

(4) We should do everything that is useful for the people. 定语从句

(5) You must work hard if you are afraid of failing. 状语从句

2.

A breed is a population of (related) animals that look alike in apparent characteristics and that also pass these apparent characteristics on to their descendants.

此句属于复合句中的定语从句。

They have defined several conditions (often appearance characteristics) that an animal must comply with.

此句属于复合句中的定语从句。

译文：

品种有许多不同的定义，但它们有一些共同的特征：一个品种是一个（相关的）动物的种群，它们在表面特征上看起来很相似，并且能把这些明显的特征传给它们的后代。一个品种的动物与另一个品种的动物不同。

品种通常是通过系谱登记部门或品种协会建立起来的。他们制定了动物必须具备的几个条件（通常是外表特征）。这些条件就称为品种标准。

3.

(1) Sex chromosome refers to the chromosome that carries a sex gene.

(2) Gene is DNA fragment which has genetic effect.

(3) Holstein is a world famous dairy breed for its high milk production.

(4) Jersey is the breed of dairy cow which Mr. White wants to bring in.

(5) I thought this swine had been vaccinated.

Extra Reading

1. B 2. D 3. A 4. D 5. D 6. C 7. B 8. C 9. D 10. C

Writing Practice

For a dairy farm, to get the cow with high milk production and strong hoof is quite important. In order to achieve this goal, we have three works to do. 1st, choose the healthy parents who contain this two characteristics. We have to buy the healthy bull's sperm. 2nd, we have to invite a breeder who can operate artificial insemination to the cows. 3rd, we have to provide good condition for the cows, and use the proper method to feed the cows.

听力材料

Part A

W: Hey Wilson, what are you going to do this weekend?

M: Hi Anna. Still working at the laboratory!

W: So what exactly do you do at the laboratory?

M: For this weekend, I am going to check the bull semen quality.

W: That sounds interesting. I haven't done anything like that. Do you mind if I go there and have a look?

M: No, of course not. Call me when you come.

W: OK, thanks a lot. Maybe I can give you a hand.

M: Well... maybe not.

Part B

W: Good morning, Mr. Zhang.

M: Good morning, everyone. Today we are going to do chicken slaughter measurements. Who can share with us how many indexes we should make?

W: I can. I think there should be 6 indexes.

M: And what are they?

W: Live-weight, carcass weight, semi-eviscerated weight, eviscerated weight, breast meat rate, and leg meat rate.

M: OK, good. Do you have anything else to add?

W: Oh... sorry. I missed one, abdominal fat percentage!

M: Yes, excellent! Now we will work in groups of four to finish this class.

W: OK.

M: There will be a total of 8 teams, each team should hand in your lab report after class.

W: No problem.

Unit 5

Warming up

1. A. cowshed B. henhouse C. trough D. water drinking E. sheep barn F. incubator

2. An incubator is a device simulating avian incubation by keeping eggs warm and in the correct humidity, and if needed to turn them, to hatch them. Scientist invented this incubator for the eggs even if there is no hen to hatch it. In industrial incubation, there are two common used methods of incubation. In single-stage incubation, the incubator contains only eggs of the same embryonic age. The advantage of single-stage incubation is that climate conditions can be adjusted according to the needs of the growing embryo. In multi-stage incubation the setter contains eggs of different embryonic ages.

Listening and Speaking

Part A

1. (1) F (2) T (3) T (4) T (5) F

2. 详见本单元参考答案后听力材料。

3.

(1) Take charge of a new automatic egg production construction.

(2) Auto water supply, auto feed, auto egg collection. To summaries, it's all automatic.

(3) 略。

Part B

1. (1) isn't (2) more (3) South Africa (4) delicate, good

2. 例:

Ostrich meat tastes similar to lean beef and is low in fat and cholesterol, as well as high in calcium, protein and iron. Uncooked, it is dark red or cherry red, a little darker than beef. Ostrich stew is a dish prepared using common ostrich meat. In some countries, people race each other on the backs of common ostriches. The practice is common in Africa and is relatively unusual elsewhere. The ostriches are ridden in the same way as horses with special saddles, reins, and bits. However, they are harder to manage than horses.

Reading

1. 略

2.

(1) crop，livestock

(2) pesticides，fertilizers，genetically modified organisms，antibiotics and growth hormones

(3) profitable，personally rewarding

(4) harmonious

(5) critical

3. 例：

Environmentally friendly；profitable；energy efficiency；food security...

译文：

有 机 农 业

有机农业是一种农作物和畜牧业的生产方法，涉及的不仅仅是不使用杀虫剂、化肥、转基因生物、抗生素和生长激素。

有机生产是一个整体系统，旨在农业生态系统内，使不同群落的生产力和和健康达到最大化，它包括土壤生物、植物、家畜和人类。有机生产的主要目标是发展可持续的并与环境协调的企业。

为什么是有机农场?

农民想要有机农场的主要原因是他们对环境的关注以及对传统农业系统中农业化学品使用的担心。还有一个问题是农业中使用了大量的资源，因为许多农业化学品要求经过能源密集的制造工艺生产，这些工艺严重依赖于化石燃料。有有机意识的农民则发现他们的耕作方法是有利可图的而且对人类是有益的。

为什么买有机产品?

消费者购买有机食品有许多不同的原因。许多人想购买不含化学杀虫剂或没有使用常规肥料种植的食品。有些人只是喜欢尝试新的不同的产品。产品口味、对环境的关注以及避免转基因食品是许多消费者喜欢购买有机食品的原因。在2007年，据估计超过60%的消费者购买了有机产品。约有5%的消费者被认为是核心有机消费者，他们购买有机食品的比例占总数的50%以上。

成功的有机农业

在有机生产中，农民选择不使用那些其他农民使用的极为方便的化学工具。生产系统的设计和管理是有机农场成功的关键。选择互补的企业，以及作物轮作以避免或减少作物问题。

根据管理的成功程度，每个有机作物的产量各不相同。从传统农业到有机农业的过渡过程中，有机农业产量低于传统水平。但经过3至5年的过渡期，有机农业的产量通常会随时间增加。

有机农业对农民来说是一种可行的备选的生产方法，但也存在诸多挑战。有机农业成功的关键之一是对使用有机方法来解决生产问题持开放态度。弄清问题的根源，评估策略以避免或减少长期的问题，而不仅仅是提供一个短期的解决方案。

Grammar

1.

(1) It is true that smoking can cause cancer.

(2) It is a serious matter that he has been late for work over and over.

(3) It is a fact that the world is round.

(4) It has been known for years that the seas are being overfished.

(5) It is a trend that organic food is becoming more and more popular.

2. (1) C (2) A (3) B (4) D (5) A

3.

(1) What people want to eat is fresh food that has never been contaminated.

(2) It is said that 10,000 cows were introduced to their new pastures last year.

(3) Whether they agree with our visit to the farm is a problem.

(4) Where we should keep the pigs has not been decided.

(5) It is true that poultry farming is profitable.

Extra Reading

1. A 2. A 3. C 4. B 5. A 6. C 7. D 8. C 9. C

Writing Practice

1

A. 例：

Fresh; sterile; delicious; good quality; pollution-free and so on.

B. 例：

Flesh air, flesher pork.

Yu tong brand, always give you the best choice.

听力材料

Part A

W: Hey Jackson, long time no see, how are things going with you recently?

M: Hi, Gucci. I am fine, thank you. Actually, I got a promotion on the farm last month.

W: Really? Congratulations! So, tell me about your new position.

M: Well, they let me take charge of a new automatic egg production construction.

W: A new automatic egg production construction? Can you tell me more details?

M：Auto water supply，auto feed，auto egg collection. To summarize，it's all automatic.

W：That is so impressive! It must cost a lot of money.

M：I heard it might be worth as much as 400，000 yuan，but it's well worth it.

W：I believe it. All the best with your new job!

Part B

W：Hey Jim，do you know what an ostrich is?

M：Yes，of course. It's a kind of wild animal，isn't it?

W：No. Actually，ostriches have been farmed in South Africa since 1860.

M：Really? What for?

W：In South Africa their feathers were used for tribal ceremonial dress and were also exported to Europe and America where they were made into ladies' fans and used for decorating hats.

M：That is interesting!

W：Yes，and it can be treated to produce about half a square meter of leather --very delicate，fine stuff of very good quality.

M：Oh，I remember the ostrich could also be as a food source.

W：Yes. Ostrich meat is slightly higher in protein than beef and much lower in fats and cholesterol. It tastes good too.

M：Amazing! What a wonderful animal!

Unit 6

Warming up

1. A. pellets　B. vegetable　C. seeds　D. hey

2. E. animal wellbeing　F. milking　G. feed　H. housing　I. animal disease J. working routines

Listening and Speaking

Part A

1. (1) T　(2) T　(3) F　(4) T　(5) F

2. 详见本单元参考答案后听力材料。

3.

(1) The boiled beef in chili soup.

(2) 略。

Part B

1. (1) F　(2) F　(3) T　(4) F　(5) T

2. 略。

Reading

1. 略。

2. (1) Healthy　(2) udder healthy, calving ease　(3) energy, proteins, vitamins　(4) calving　(5) vulva, hands

3. 例：

Whenever you need to feed the animals, you must pay attention to three points. First, you should know the anatomy of the animal you are feeding, understand the habits of the animals and give it the correct nutrients. Second, pay close attention to the hygiene of the environment in which the animals live (this is quite important for the disease prevention). Third, pay attention to the signals the animal sends you which can help you focus on the animal health or direct your attention to something you take care.

译文：

奶牛的科学饲养

奶牛的科学饲养重点集中在三个方面。

第一：繁殖力。在奶牛场，奶牛的健康和繁殖力是非常重要的。健康的奶牛产更多的奶，更多产，也活得更久。繁殖力高的母牛产犊间隔期短，不需要多次输精就能怀孕。公牛可以将其健康体魄和繁殖力传给其后代。如果育种值超过100，其后代将会更多产、更健康，等等。育种值的几个指标包括：乳房健康、繁殖力、生命力和产犊顺畅。

第二：营养。好的营养意味着提供充足的能量、蛋白质、矿物质和维生素。提供平衡的日粮配比不仅可使母牛产奶量提高，而且也会使其有更好的繁育表现。在泌乳初期，当产奶量达到高峰时，每天的干物质摄取量很难适应（满足）奶牛的营养需求，特别是高产奶牛。

母牛的干物质摄取在哺乳早期进展缓慢，因此在这个时候，每天的能量不足是很常见的。如果母牛的日粮没有足够的绿粗饲料或高水平的副产品饲料，就可能导致维生素A、磷、铜、钴、碘和/或硒的缺乏。这可能引起高产奶牛的健康问题。母牛持续摄入所需成分的高品质的矿物质是很重要的。

饲料配比中有充足的高质量的粗饲料，按配方提供正确水平的蛋白质、能量、矿物质、维生素和微量元素将会使产犊和产后初次发情的时期缩短。

第三：卫生。好的卫生，特别在产犊前后是至关重要的。在产犊之前清洁母牛的阴部、产绳和你的双手，为母牛产犊准备一个干净且消过毒的产房将就可以了。

如果忽视这些因素，可能会引起子宫炎症。它会影响母牛随后的生育能力，还会延长子宫为下一次怀孕需要恢复的时间。子宫内膜炎可以通过从外阴排出的白色黏液来诊断。兽医可以进行治疗，另一方面，当母牛又开始发情时，子宫也可以自然地清洁自身。

Grammar

1.

(1) Everyone knew what happened and that she was worried. 宾语从句

(2) The reason lies in that she works harder than the others do. 宾语从句

(3) Let's see whether we can find out some information about Jersey cows. 宾语从句

(4) I asked hin where I could get that much money. 宾语从句

(5) The problem is that the water on our farm is not clean. 表语从句

2.

If you know how an animal behaves in a specific situation you can take this into account.

此句属于复合句中的宾语从句。

Coordinating human behavior with animal behavior, this is what is involved in animal handling.

此句属于复合句中的表语从句。

译文：只有当一个穿着白大褂的兽医站在动物面前时，动物才能在这一刻意识到什么事情在发生。此时你怎样使用特殊的应对办法呢？如果你知道动物在特殊的情况下如何反应，那就可以采用相应方法应对。当人们接近动物时，动物就会对人做出反应。

根据动物的行为来协调人的行为，这些都属于动物行为操控的内容。

3.

(1) That's why the production of the dairy cow decreased.

(2) The problem is whether we can find the cause of the disease.

(3) Do you know the importance of the calcium salt to the animals?

(4) The animals may reduce their productivity when they are stressed suddenly.

(5) The dairy market now is different from the situation a few years ago.

Extra Reading

1. A 2. D 3. B 4. B 5. A 6. A 7. B 8. C 9. D 10. C

Writing Practice

When feeding your chickens, there are four points you have to learn. First, 5 Essential Dietary Ingredients, which including meat protein, grass and hay, dried whole corn and grains, greens, calcium. Second, feeding your hens eggshells. Third, providing good quality feed is essential for both chicken health and maximum egg yield. Fourth, treats are nice. A chicken' s diet dictates their overall health and their egg production. Never skimp on feeding your flock. Keep their food clean, dry and vermin free by storing in sealed containers. And finally, fresh water is as important as quality food.

听力材料

Part A

W: How do you like the boiled beef in chili soup?

M: I love the tender beef, but it' s far too spicy. I can' t feel my tongue!

W: No wonder you only had a little bit.

M: Sichuan food is different from food in the US, right?

W: Right. The overseas Chinese food isn' t as authentic as real Chinese food.

M: Having Sichuan food in China is like a taste adventure. I' ll tell my tongue to be careful.

W: We may need to get something else now. It seems like you are still hungry.

M: Thank you. Some bread will do.

Part B

W: Hello! I am doctor Wang. What can I do for you?

M: Hi, I'm Tom. I've got a new dog. This is my first time to have a pet dog. I don't know what to feed him.

W: Don't worry. We're here to help find the right food for your pet.

What's your dog's breed?

M: He is a Golden Retriever puppy.

W: What's your pet's age? Puppies, adults and seniors all have different and special nutritional needs.

M: Ummm, he is 8 months old.

W: Does your pet have any special health considerations?

M: Oh, I have no idea about that.

W: OK, we have a dog food with a formula that is designed for the specific needs of Golden Retriever puppies. Digestive health is important during the puppy's growth period. Their digestive system is not yet fully developed and cannot absorb nutrients in the same way as adult dogs.

M: That sounds good, I will choose this one.

W: Be careful not to overfeed your puppy. Always follow the instructions on the package.

Unit 7

Warming up

1.

(1) Foot-and-mouth disease (2) Blue-ear pig disease

(3) Bluetongue (4) The bird flu/Avian influenza

(5) Duck pestilence (6) Foot-and-mouth disease

(7) Avian pox (8) Edema disease of pig

(9) Deformity caused by a deficiency of riboflavin

2.

(1) — e (2) — c (3) — b (4) — a

(5) — d (6) — f (7) — h (8) — g

Listening and Speaking

Part A

1. (1) T (2) F (3) F (4) T (5) T

2. 详见本单元参考答案后听力材料。

3.

(1) **The common poultry diseases include the followings:**

Avian Influenza 禽流感: Avian influenza is caused by several viruses. It affects mainly turkeys and chickens. The birds cough, sneeze, and lose weight. Death rate is low. Good management helps in prevention. Treatment with antibiotics may help reduce losses.

Avian Pox (Fowl Pox) 禽痘: Avian pox is caused by a virus and affects turkeys, chickens, and other birds. Symptoms include wart-like scabs around the head and comb yellow cankers in the mouth and eyes, and, in turkeys, yellowish-white cankers in the throat. Young birds have a slower rate of growth. Egg production in layers is reduced. The disease is prevented by vaccination. There is no treatment.

Duck Virus Enteritis (Duck Plague) 鸭病毒性肠炎 (鸭瘟疫): Duck virus enteritis is caused by a virus. It affects ducks, geese, and other waterfowl. Symptoms include watery diarrhea, nasal discharge, and droopiness. Death loss is high. Sanitation and isolation of birds helps to prevent the disease. There is no treatment.

(2) Prevention involves sanitation, good management and vaccination.

Part B

1.

(1) It is one subtype of influenza viruses that is sometimes found in birds or poultry.

(2) Yes.

(3) By close contact with infected birds or environments contaminated with the flu virus.

(4) No，right now there is no vaccine to protect against this virus.

(5) Firstly，avoid contacting with infected poultry and birds. Secondly，clean your hands thoroughly with water immediately after contacting with infected poultry and birds.

2. 略

Reading

1. 略

2.

(1) blue-ear pig disease/ PRRS

(2) discolored ears

(3) late-term fetal death

(4) the United States

(5) pigs

3. 例：

Swine edema is caused by toxins produced by a bacterial infection. Pigs between 3 and 14 weeks of age are most commonly affected. Sudden death may be the first symptom seen. The disease may be confused with chemical poisoning. Other symptoms include refusal to eat，convulsions，staggering，and swollen eyelids. There is usually no fever.

Swine fever：All the year round，any pigs may be affected and spread quickly. The symptoms of the disease can include high fever，purulent conjunctivitis，constipation，and diarrhea. There is also the skin's bleeding in ears，neck，abdomen and inside of the limbs，which does not fade even after applying pressure.

译文：

猪 蓝 耳 病

猪呼吸与繁殖综合征病毒是一种能使猪患“猪繁殖和呼吸障碍综合征(PRRS)”的病毒，这种病也被称为猪蓝耳病，病名来源于感染了病毒的猪耳朵可能会出现暂时变色，这是当今养猪行业中影响经济效益最严重的疾病之一。PRRS是一种全球性的毁灭性猪病。1987年，该病首次在美国报道，三年后在西欧出现并迅速传播。该病导致美国养猪行业每年损失约6.44亿美元，欧洲最近估计每年损失近15亿欧元。

PRRS的症状表现包括孕晚期流产、产死胎、弱仔和仔猪患严重呼吸系统疾病。它曾一度被称为“神秘猪病”，现在人们才知道该病是由某种病毒引起的，感染了病毒的猪更易发生细菌感染。尽管已有疫苗可用，但目前对PRRS还无法治疗。已经接种过疫苗的猪群仍有暴发疫情的可能。现已研发出一种基因测试，它可以区分疫苗中所发现的病毒无害毒株和真正的致病病毒。遵循良好的生物安全管理措施等将有助于降低养殖场中PRRS的发病率。

世界动物卫生组织宣称：猪蓝耳病主要发生在全球的生猪养殖主区域。例如，一场蓝耳猪病的暴发曾在一年内导致中国多达100万头猪死亡。中国政府表示，通过接种疫苗和宰杀受感染的猪，他们在控制蓝耳病传播方面取得了重大进展。

专家们表示猪繁殖和呼吸障碍综合征（PRRS）是一种复杂的疾病，控制它的主要措施是改良活疫苗（MLV）。他们还证实，这种疾病似乎不会影响猪以外的其他动物，而且目前还未发现人类感染此病的病例。

Grammar

1.

（1）定语从句

（2）同位语从句

（3）定语从句

（4）定语从句

（5）同位语从句

2.

Rinderpest（牛瘟）can spread quickly through the air and in water which contains waste from animals with the virus.

Rinderpest is only the second disease that was ever declared to have been eliminated.

译文：

牛瘟可以通过空气和被携带病毒的动物粪便污染的水进行传播。该病死亡率可达80%～90%。本病主要感染牛和水牛，还有长颈鹿、牦牛、羚羊等其他动物。

目前，牛瘟专家约翰·安德森（John Anderson）称，该疾病的终结是“兽医史上最伟大的成就。”官方表示，他们决定仍保留一些病毒样本和受感染的组织以便今后研究。牛瘟是被宣布已消灭的第二种疾病，另一种是天花。

3.

（1）The disease does not seem to affect animals other than pigs.

（2）The blue-ear pig disease was first recognized in the United States.

（3）She is a veterinary epidemiologist——an expert in the spread of diseases involving animals.

(4) The name for the virus comes from the fact that infected pigs can temporarily develop discolored ears.

(5) Pigs that have been weakened by the virus are more likely to get bacterial infections.

Extra Reading

1. C 2. B 3. C 4. A 5. D 6. D 7. C 8. A 9. B 10. D

Writing Practice

Ladies and Gentlemen,

Welcome to my chicken farm, I am the farm owner, Henry Harrison. My farm was built in 2010 and there are about 160,000 chicken raised here in all. In order to keep the poultry houses clean and disinfected, the following rules should be followed before your visiting:

Firstly, wear clean coveralls and disinfected rubber footwear before entering the poultry area.

Secondly, slake your rubber footwear in a foot pan with disinfectant at the door.

Thirdly, during your visit, don't touch any chicks so as to prevent the chicks from getting infected.

I hope you enjoy your time here. Thank you.

听力材料

Part A

W: Welcome to my chicken farm, Mr. Tree.

M: Thank you, Mrs. Wang. I'm glad to visit your farm. Your farm looks quite big, how many chickens do you raise?

W: My farm was built in 2010. All in all, about 160,000 chickens are being raised.

M: Well, I have heard that poultry diseases break out much more easily in big chicken farms. You must have good measures in place to control these diseases.

W: Yes. The best way to control poultry diseases is prevention. Prevention involves sanitation, good management and vaccination.

M: Oh, I see. Thank you!

W: You're welcome.

Part B

W：Hello，Mr. Smith. May I ask you some common questions about the new virus，H7N9?

M：Yes. Go ahead.

W：Well，what is H7N9?

M：It is one subtype of the influenza virus that is sometimes found in birds or poultry.

W：Can the virus H7N9 infect humans?

M：Yes. China reported more than 126 cases of human infection with the H7N9 virus in 2013.

W：How do people get infected with the flu virus?

M：Human infections with the flu virus are usually after close contact with infected birds or environments contaminated with the flu virus.

W：Is there a vaccine to protect against this new H7N9 virus?

M：No，right now there is no vaccine to protect against this virus.

W：What basic measures should people take to prevent getting infected with this flu virus?

M：Firstly，avoid contact with infected poultry and birds. Secondly，clean your hands thoroughly with water immediately after contact with infected poultry and birds.

W：Oh，I see. Thank you so much.

M：You are welcome.

Unit 8

Warming up

1.

(1) Brucellosis
(2) Bird flu/ Avian influenza
(3) Rabies
(4) Mad cow disease
(5) Anthrax of cattle
(6) Swine streptococcal diseases
(7) Cutaneous Leishmaniosis
(8) Yersinia pestis
(9) Hydatid disease of liver

2.

(1) —d (2) — e (3) — a (4) — f (5) — b
(6) — c (7) — i (8) — g (9) — h

Listening and Speaking

Part A

1. (1) F (2) F (3) T (4) T (5) T

2. 详见本单元参考答案后听力材料。

3.

(1) **Rabies**: Hydrophobia (fear of water) is the historic name for **rabies.** It refers to a set of symptoms in the later stages of an infection in which the person has difficulty swallowing, shows panic when presented with liquids to drink, and cannot quench his or her thirst.

(2) Firstly, wash the cuts under cold running water for 20～30 minutes as soon as possible.

Secondly, go to the hospital or a Disease Control and Prevention Center to take the necessary treatment, for example, getting vaccinated.

Part B

1.

(1) Rabies, anthrax, foot-and-mouth disease, bird flu and mad cow disease.

(2) Anthrax.

(3) Anthrax enters the human body by breathing, eating, or through broken skin.

(4) To avoid contact with infected animals as much as possible and avoid eating raw or undercooked meat.

(5) Vaccination and preventative antibiotics

2. 略

Reading

1. 略

2.

(1) Two million people.

(2) Zoonosis.

(3) More than sixty.

(4) Three (from food; direct contact with animals; through water or the air).

(5) Not only to prevent and treat diseases but to increase growth.

3. 例:

(1) **Common zoonoses**: rabies, foot-and-mouth disease, mad cow disease, avian influenza, tuberculosis.

(2) **Rabies transmission**: The most common mode of rabies virus transmission is through the bite and virus-containing saliva of an infected host. Though transmission has been rarely documented via other routes such as contamination of mucous membranes (i. e., eyes, nose, mouth), aerosol transmission, and corneal and organ transplantations.

译文:

人畜共患传染病

研究者估计，每年有超过20亿人感染上动物传播的疾病，超过200万人因此死亡。这类在人与动物之间传播的疾病称为人畜共患传染病。

人类大多数疾病实际上都是人畜共患的疾病，超过60%的人类疾病是由于其他脊椎动物传播感染的。其中一些疾病很常见，如一些食源性疾病以及肺结核、细螺旋体病等是很常见的，而其他类的疾病是很罕见的。

当前主要的几种疾病都是人畜共患病，如埃博拉病毒病、沙门氏菌病和禽流感。尽管艾滋病现在已经进化为人类独有的一种疾病，但它是在20世纪早期传染给人类的一种人畜共患疾病。人畜共患疾病可以由一系列疾病病原体引起，如病毒、细菌、真菌和寄生虫。在已知可感染人类的1 415种病原体中，61%是人畜共患的。人类的大多数疾病起源于动物；但是只有涉及动物向人类经常传播的疾病，如狂犬病，才被认为是直接的人畜共患病。

人畜共患病有不同的传染方式。直接传染就是直接通过如空气（禽流感）、叮咬和唾液（狂犬病）从动物传染给人类的疾病。间接感染则是通过一个携带疾病病原体的中间物种（称为载体）传染疾病。当人类传染给动物时，则称为反向人畜共患病或人类传染病。

对于人来说可以有多种感染途径，或许最常见的就是病从口入，其他传播途径包括与动物直接接触，还有一些疾病通过水或空气传染。

随着全球人口不断增长，对肉产量的需求也不断增加，未来几年情况会更加严峻。高产的农场通常采用密集型的饲养方式，而拥挤的饲养环境会使疾病传播更为迅速。另一个令人担心的就是在食品动物的养殖中抗生素的滥用，现在使用抗生素不仅仅是为了预防和治疗疾病，更多地是为了促生长。

Grammar

1.

(1) You will find it where you left it. 地点状语从句

(2) To search for gold, many people went to California. 动词不定式作目的状语

(3) It is very kind of you. 副词作状语修饰形容词

(4) Please speak politely. 副词作状语修饰动词

(5) Working in this way they greatly cut the cost. 现在分词作状语

2.

(1) Where there is a will, there is a way. 地点状语从句

(2) Everybody likes him as he is kind. 原因状语从句

(3) Sorry，I was out when you called me. 时间状语从句

(4) He is ill，so he can't go to school. 结果状语从句

(5) Whatever you say，I won't believe you. 让步状语从句

3.

If swine flu becomes a problem，it may be wise to use a vaccination program to help control the outbreak.

If 引导的条件状语从句，表示"如果......，那么......"。

After the disease is brought under control，only replacement gilts need to be vaccinated.

After 引导的时间状语从句，表示"在......之后"。

If vaccinating the sows does not control the disease，the pigs should be vaccinated when they are 7 to 8 weeks of age.

If 引导的条件状语从句，表示"如果......，那么......"；

When 引导的时间状语从句，表示"当......时候"。

译文：

猪流感是一种由病毒和细菌共同作用而引起的呼吸道疾病。症状包括：发热、呼吸困难、咳嗽、食欲不良和虚弱。猪发病突然，但通常在 6 天后即可康复。

如果猪流感较为严重，那么最佳的办法就是通过接种疫苗来控制疫情的暴发。一般说来，通过对母猪每年接种两次疫苗即可控制该病。在疾病得到控制后，只需对后备母猪接种疫苗即可。如果对母猪接种疫苗还无法控制病情，那么当仔猪达到 7 至 8 周龄时就应该接种疫苗。

Extra Reading

1. A 2. D 3. C 4. C 5. B 6. C 7. B 8. D 9. A 10. D

Writing Practice

Rabies is a viral disease that causes inflammation of the brain in humans and other mammals. Rabies causes about 24，000 to 60，000 deaths worldwide per year. So it is very important for us to know how to reduce the risk of getting infected with rabies.

The following methods can help to reduce the risk of contracting rabies：

• Vaccinating dogs and cats against rabies.

• Keeping pets under supervision.

• Not handling wild animals or strays.

• Contacting an animal control officer upon observing a wild animal or a stray，especially if the animal is acting strangely.

• If bitten by an animal，washing the wound with soap and water for 10 to 15 minutes and contacting a healthcare provider to determine if post-exposure prophylaxis is required.

听力材料

Part A

M：What's the matter? You don't look well.

W：Oh，doctor. Just now，I was playing with my pet puppy in the garden，but it suddenly scratched my leg，which scared me. Look，here are two shallow wounds on my leg.

M：Let me have a look. Oh，luckily，it looks like only the skin is cut.

W：Do I need to take any treatment?

M：Yes，of course. First of all，you have to wash the wounds under cold running water for 20 to 30 minutes as soon as possible.

W：Okay，I did that just now.

M：Well done. Then you have to get injected with a vaccine for the prevention of rabies. You should get injections of the vaccine on day 1，day 3，day 7，day 14 and day 30，five times in all.

W：I see. Thank you very much，doctor.

Part B

W：Hello，Professor Smith. May I ask you something about the zoonoses?

M：Yes，of course.

W：What do the common zoonoses include?

M：Rabies，anthrax，foot-and-mouth disease，bird flu and mad cow disease are the most common zoonoses.

W：Oh，I see. And I often hear about anthrax in my daily life. Would you please tell me how anthrax transmits from animals to humans?

M：Well，anthrax can be spread by contact with the spores of the bacteria，which are often from infectious animal products. Anthrax can enter the human body by breathing，eating，or through broken skin.

W：So，what prevention measures should we take in daily life?

M：The best way to reduce the chance of an anthrax infection is to avoid contact with infected animals as much as possible and avoid eating raw or undercooked meat. Vaccination and preventative antibiotics are also the good preventive measures against anthrax.

W：OK，I think I know what to do. Thank you very much，Professor Smith.

Unit 9

Warming up

1. A—（1） B—（7） C—（4） D—（2） E—（6） F—（1） G—（5） H—（8）

2. A. more snake B. snail tortoise C. spider D. parrot E. silkworm

Listening and Speaking

Part A

1. （1）T （2）F （3）F （4）F （5）T

2. 详见本单元参考答案后听力材料。

3. （1）Gold fish，silkworm，little puppy and rabbit. （2）略

Part B

1.

The benefits of keeping pets	Pets can accompany us when we feel lonely. Pets are good for people's health. Pets are helpful.
The disadvantages of keeping pets	Pets cost a lot of money and time. Pets can transmit diseases. The noise and dung of pets are sources of pollution. Pets can also be dangerous.

2. 略

Reading

1. 略

2.

（1）a person's company，protection，and/or entertainment

（2）animal pets

（3）caregivers or pet owners

（4）clean and quiet

（5）video game

3.

Animal pets，plant pets and virtual pets are mentioned in the passage.

Animal pet，besides dogs and cats. It contains rodent pets such as gerbils，hamsters，chinchillas，fancy rats and guinea pigs，avian pets such as canaries，parakeets, corvids, parrots and chickens, reptile pets such as turtles, lizards and snakes, aquatic pets such as goldfish, tropical fish and arthropod pets.

译文：

什么是宠物？

经济的快速发展极大地改变了我们的生活方式。对于很多人而言，养宠物已成了一种新的流行爱好。但是，宠物是什么呢？通常，人们饲养宠物或动物伙伴主要是为了陪伴、保护和/或娱乐，而不是为了役用、运动、作为家畜或实验用动物。受欢迎的动物常常以它们漂亮的外表或忠诚或顽皮的个性而闻名。除了动物宠物，当今，有些植物也被当作宠物种植，甚至一些虚拟宠物也受到年轻人的追捧。

动物宠物种类繁多。大体上，最流行的动物宠物是狗和猫。但是有些人也把啮齿动物，如沙土鼠、仓鼠、南美洲栗鼠（龙猫）、花鼠和豚鼠作为他们的宠物。金丝雀、鹦鹉、鱼、鹦鹉和鸡通常也被当作宠物饲养，它们属于鸟类宠物。爬行动物，如海龟、蜥蜴和蛇在花鸟市场上并不陌生。水生宠物，如金鱼、热带鱼，以及某些节肢动物是常见的宠物。

动物宠物可能有激励它们的照顾者的能力，特别是老年人，使他们感觉这些宠物也需要被照顾，需要一起运动，帮助他们解决身体或心理方面的问题。动物的陪伴能帮助存在焦虑或抑郁情绪的人们保持可接受的快乐程度。养宠物还可以帮助人们达到健康的目标。有证据表明，养宠物有助于人们过上更长、更健康的生活。养宠物可以显著降低老年人的甘油三酯，从而降低老年人患心脏病的风险。美国国家卫生研究院的一项研究发现，养狗的人比没有养狗的人死于心脏病的几率更低。有证据表明，宠物对痴呆症患者可能有治疗作用。其他研究表明，对老年人来说，健康可能是养宠物的必要条件，而不是结果。

然而，并不是所有的人都喜欢养宠物。植物被越来越多的人视为宠物，因为他们更喜欢保持一个干净安静的环境。与宠物动物相比，宠物植物不会产生噪声和排泄物。此外，虚拟宠物在年轻人中很受欢迎。什么是虚拟宠物？实际上，它是一个电子游戏。你可以根据你的需要从一些特定的网站下载，然后安装在你的电脑或手机上。你可以给你的宠物起一个名字，并把它当作一个真正的宠物。你需要在打开电脑时用游戏开发者购买的虚拟食物来喂养它。否则，它就会消失。然而，虚拟宠物是无法取代真正的宠物的。毕竟，这是游戏开发者赚钱的方式之一。

Grammar

1. 略

2. 译文：

蛙肾的组织结构观察

高登慧[①] 李朝波[②]

（① 贵州大学动物科学学院 贵阳 550025；

② 广西农业职业学院 南宁 530001）

摘要：应用组织学和组织化学方法研究了黑斑蛙肾的显微结构。结果表明：蛙的肾除具有肾单位和集合小管外，还见有淋巴样组织分散于肾实质中。在肾腹侧发现有与真骨鱼类斯坦尼斯小体相似的结构，其中聚集有较多的肥大细胞。说明黑斑蛙的肾具有多种生理功能。

关键词：黑斑蛙；肾；组织结构；肥大细胞

3.

(1) A Sociological Study on the Relationship between People and Pets.

(2) (College of Animal Medicine, Guizhou Agriculture University, Guiyang, 550001 China)

(3) The diagnosis and treatment for common skin disease of dogs and cats were introduced in this passage.

(4) The distribution of mast cells in rabbit kidney was observed by histological and histochemical ways.

(5) The results showed that the incidence of acute pancreatitis was related to the sex, age and variety of dogs.

Extra Reading

1. A 2. D 3. A 4. B 5. B 6. A 7. D 8. B 9. D 10. B

Writing Practice

Research on the Status and Development Prospect of Pet Nutrition

ZHANG Yi-ping

(Shandong Vocational Animal Science and Veterinary Collage, Weifang 261061, China)

Abstract: Pets, also known as companion animals, are usually regarded as one of family members. With the improving of people's living, the owners are paying more attention to how to feed their pet dogs and pet cats scientifically. Raising of pets is gradually developing into the direction of commercialization. Pet foods appeared accompanying the development of economy. Nowadays, more and more people would like to raise a pet, but few people know more about pet nutrition. The researches on the status of pet feed in recent years are reviewed in this paper so as to provide reference for the study of pet nutrition.

Keywords: Pet; Pet Nutrition; Pet Foods; Pet Feed

Part A

Son: Mom, many of my classmates have a pet. I want one too.

Mom: That' s a good idea. But what kind of pet do you want to get?

Son: I would like to hear your suggestions.

Mom: How about a goldfish? It is so easy to raise.

Son：Goldfish are too boring. I want to raise a silkworm.

Mom：No! Silkworms look so scary. I'm too scared to take care of it for you when you go to school.

Son：Mom，silkworms are so interesting. And it will make me very popular to have an exotic pet.

Mom：Let's get a nice pet like a puppy.

Son：A puppy? I don't like dogs. Dogs aren't friendly sometimes.

Mom：Well，how about a rabbit? Rabbits are very gentle and have a cute pair of red eyes.

Son：Yeah，it's perfect!

Mom：Let's get one this weekend.

Son：OK，thanks mom.

Part B

J：Hey，Lily，good to see you.

L：Hi，Jerry，long time no see.

J：You look so down，what's up?

L：My dog died yesterday.

J：Oh! I am really sorry to hear that. Do you think it is a good idea to keep a pet?

L：Absolutely. They can accompany us like a friend when we feel lonely.

J：But I think it always makes for so much trouble. People have to take care of them. They're just like human babies. It costs a lot of money and time.

L：Yeah，usually I treat my dogs like my kids. However，pets are good for people's health. Some research shows dog and cat owners have less heart trouble than other people.

J：Maybe you are right，but pets can also be dangerous. They can bite people and transmit diseases.

L：But now more and more people keep pets because people often feel they are very friendly and warm. What's more，some pets are helpful. A dog can guard the door，and a cat can catch mice.

J：But what's worse than the noises and dung of pets! They are sources of pollution.

L：Okay，okay，don't say that any more. Every coin has two sides. As far as I'm concerned，we get more benefits than drawbacks from keeping pets.

Unit 10

Warming up

1. Meat products：1，7

   Dairy products：2，6，9

   Egg products：3，4，5，8

2.

Blood	eating，nutritive and medicinal value
Internal organs	eating，nutritive and medicinal value
Fur	eating，leather clothing，shoes，bags，and gloves，specimens，clothing，blankets，glue
Hoof	eating，maintaining beauty and keeping young，nutritive and medicinal value
Bone	nutritive and medicinal value

Listening and Speaking

Part A

1. (1) F (2) F (3) F (4) T (5) F
2. 详见本单元参考答案后听力材料。
3. 略

Part B

1. Keys：Milk，nutrients，protein，vitamins，Cows，goats，sheep，water buffalo，Asia，horses，camels
2. 略

Reading

译文：

肉　产　品

肉是最有价值的畜产品。肉的成分除了蛋白质、氨基酸、矿物质、脂肪、脂肪酸和维生素外，还包括其他类生物活性成分和少量的糖。从营养学角度看，肉类的重要性体现在以下几个方面：一，具有高品质蛋白质和氨基酸；二，矿物质和维生素的生物活性高。

发达国家肉类的消费一直相对稳定，但自 1980 年以来，发展中国家每年人均肉类消费量翻了一番。随着人口的增长和人民收入水平的提高，人们对食物的喜好也发生了改变，对畜产品的需求也不断增加。

据预测，到2050年，世界肉类产量将翻一番，而大部分的增长将发生在发展中国家。这些国家肉产品销售的不断扩大为养殖从业人员和肉品加工商提供了重要机会。然而，伴随畜禽产品的不断增加，肉类加工、销售、质量和卫生安全也面临严峻问题。

国际粮农组织通过规范肉类产品的安全、高效和可持续生产，为其成员的畜牧业发展和扶贫提供机会。

本计划的重点是改进并提升小规模肉类生产和加工的技术，提高小规模生产者的技能和能力。粮农组织通过总部与实地活动相结合，部分国家、地区和国际合作伙伴协作的方式，帮助销售肉产品和完善肉产品价值链。

1. 略

2.

(1) Meat is composed of protein and amino acids, minerals, fats and fatty acids, vitamins, other bioactive components, and small quantity of carbohydrates.

(2) From the nutritional point of view, the importance of meat is derived from its high quality (in) protein, amino acids and its highly bioavailable minerals and vitamins.

(3) Meat consumption has been relatively stable in the developed world, annual per capita consumption of meat has doubled since 1980 in developing countries.

(4) It is also a great challenge for increasing the production of livestock, safely processing and selling hygienic meat and meat products.

(5) By regulating the safe, efficient and sustainable production of meat products, the The Food and Agriculture Organization (FAO) aims to provide opportunities for livestock development and poverty alleviation in its member states.

3. 例：

Meat products refer to raw materials used in livestock and poultry meat, which are all processed meat or semi-finished products, such as sausage, ham, bacon, barbecue meat, etc.

Grammar

1.

(1) the content of research

(2) the objective of research

(3) the result of research

(4) the result of research

(5) the conclusion of research

2. 译文：

摘要：本研究旨在研究鸭脂肪对鸭风味的影响。用石油醚和氯仿与甲醇分别选择性提取鸭肉中的甘油三酯和磷脂。感官分析和仪器分析表明，甘油三酯对鸭肉香气的影响不大，而磷脂则相反。用甲醇萃取氯仿提取的鸭脂，水洗去除脂肪中的水溶性化合物。洗涤鸭脂肪中含有 NO、S-和 N 类化合物，其特征香气成分显著低于对照鸭脂（$p<0.05$）。鸭脂肪中的水溶性物质对鸭肉香气有作用，但对鸭肉香气的作用不明显。

3.

(1) The egg quality of the Sichuan goose was studied in this paper.

(2) This research attempted to study the effect of Chinese herbal medicine on the content of cholesterol in quail egg yolk.

(3) The sensory properties of chicken meat balls prepared under different conditions was studied through orthogonal experiment.

(4) The results showed that gender obviously affected the attributes and quality of pork.

(5) It is concluded that the stability of goat's milk is best when the pH is 6.8.

Extra Reading

1.D 2.D 3.B 4.D 5.A 6.C 7.A 8.B 9.A 10.C

Writing Practice

略

听力材料

Part A

H：What are we going to eat for dinner?

W：I'm going to fix some steak.

H：I'm afraid the meat is rotten.

W：That's strange! I just bought it yesterday.

H：Well, I forgot to put it in the refrigerator.

W：What!? Now what should we eat?

H：Don't worry! There are still some leftover eggs and pork chops.

W：Oh my goodness! We have been eating pork for a week.

H：Do you have any other suggestions?

W：Why don't we eat out?

H：That's a good idea. Let's go.

Part B

Milk is often seen as the most complete natural food. It provides many of the nutrients for the health of the human body and is an excellent source of protein, vitamins and minerals, too. Cows are not the only source of milk, because milk can be obtained from many different sources. The milk from goats and sheep make a great contribution to the total milk production all over the world, especially in Europe. On the contrary, the water buffalo is a common source of milk in many parts of Asia while milk from horses or camels is also drunk by some people.

参考文献

Arima Y.，Vong S.，Shimada T.，2013. Western Pacific Surveillance and Response Journal [M]. Manila：WHO.

James R. Gillespie，Frank B. Flanders，2010. Modern Livestock and Poultry Production [M]. Eighth Edition，NY：Delmar Cengage Learning.

Malcolm Sue，Goodship J.，2007. Genotype to phenotype [M]. Second edition. Oxford：Academic Press.

Meter T.，2018. Egg tray for incubating and hatching eggs [M]. Veenendaal：United States Patent Application Publication.

Morgan Thomas Hunt，1919. Monographs on Experimental Biology-The Physical Basis of Heredity [M]. 2011 edition. Redditch：Read Books Press.

Robert Edwards，2007. the path to IVF [J]. Reproductive biomedicine online，23 (2) 245-62.

Skowronski DM，Janjua NZ，De Serres G，2013. Eurosurveillance [M]. Stockholm：European Centre for Disease Prevention and Control.

Taylor Peter，Lewontin Richard，2017. The Genotype-Phenotype Distinction：The Stanford Encyclopedia of Philosophy [M]. Summer 2017 Edition. Stanford：Stanford University publishing.

Thacker E，Janke B，2008. The Journal of Infectious Diseases [M]. Oxford：Oxford University Press.

Thomas O. McCracken，Robert A. Kainer，Thomas L. Spurgeon，2006. Spurgeon's Color Atlas of Large Animal Anatomy：The Essentials [M]. Iowa：Blackwell Publishing.

陈霞，2017. 鸡蛋产品鉴伪技术的构建 [J]. 食品工业科技 (1)：300-303.

董亚芬，2011. 大学英语 1 (精读) [M]. 3 版 . 上海：上海外语教育出版社 .

杜文贤，2016. 宠物专业应用英语 [M]. 北京：化学工业出版社 .

霍恩比，2002，牛津高阶英汉双解词典 [M]. 4 版 . 李北达，译 . 北京：商务印书馆，牛津大学出版社 .

焦洪超，2011. 山东省蛋鸡生产状况的调查与分析 [J]. 中国家禽，33 (1)：64-67.

雷达，2010. 喝上放心奶 国外有啥招 [J]. 质量探索，(6)：54-55.

李颖，2016. 关于牛奶的十大谣言和真相 [J]. 中国质量万里行，(12)：14-15.

马文涛，2012. 对牛奶产品安全生产的几点看法 . 中国牛业进展 [M]. 北京：中国农业科学技术出版社 .

秦秀白，2010. 新世纪大学英语系列教材综合教程 [M]. 上海：上海外语教育出版社 .

文秋芳，2010. 应用语言学：研究方法与论文写作 [M]. 北京：外语教学与研究出版社 .

闫付军，2014. 鲜活肉蛋产品免税对企业的影响 [J]. 金融经济，(16)：175-176.

尹尚远，张沙拉，2016. 英语语法分解大全 [M]. 南京：江苏凤凰科学技术出版社.
翟象俊，2016. 21世纪实用英语语法教程 [M]. 上海：复旦大学出版社.
张超，2014. 蛋产品里有大学问 [J]. 农产品加工，(1)：21.
张家骅，田文儒，李跃民，2012. 畜牧兽医专业英语 [M]. 北京：中国农业出版社.
张森，李荣丽，2015. 应用型技能英语 [M]. 北京：电子工业出版社.

图书在版编目（CIP）数据

畜牧兽医专业英语/卫金珍主编.—北京：中国农业出版社，2018.12

高等职业教育农业农村部“十三五”规划教材

ISBN 978-7-109-24579-2

Ⅰ.①畜… Ⅱ.①卫… Ⅲ.①畜牧学－英语－高等职业教育－教材②兽医学－英语－高等职业教育－教材 Ⅳ.①S81②S85

中国版本图书馆CIP数据核字（2018）第208002号

中国农业出版社出版

（北京市朝阳区麦子店街18号楼）

（邮政编码100125）

责任编辑 徐 芳

北京通州皇家印刷厂印刷 新华书店北京发行所发行

2018年12月第1版 2018年12月北京第1次印刷

开本：787mm×1092mm 1/16 印张：11

字数：225千字

定价：38.50元